Critical Mass: The One Thing You Need to Know About Green Cars

Critical Mass: The One Thing You Need to Know About Green Cars

FELIX LEACH AND NICK MOLDEN

SAE INTERNATIONAL®

Warrendale, Pennsylvania, USA

SAE INTERNATIONAL®

400 Commonwealth Drive
Warrendale, PA 15096-0001 USA
E-mail: CustomerService@sae.org
Phone: 877-606-7323 (inside USA and Canada)
724-776-4970 (outside USA)
FAX: 724-776-0790

Library of Congress Catalog Number 2024937790
http://dx.doi.org/10.4271/9781468608229

Information contained in this work has been obtained by SAE International from sources believed to be reliable. However, neither SAE International nor its authors guarantee the accuracy or completeness of any information published herein and neither SAE International nor its authors shall be responsible for any errors, omissions, or damages arising out of use of this information. This work is published with the understanding that SAE International and its authors are supplying information but are not attempting to render engineering or other professional services. If such services are required, the assistance of an appropriate professional should be sought.

ISBN-Print   978-1-4686-0821-2
ISBN-PDF   978-1-4686-0822-9
ISBN-epub 978-1-4686-0823-6

To purchase bulk quantities, please contact: SAE Customer Service

E-mail: CustomerService@sae.org
Phone: 877-606-7323 (inside USA and Canada)
724-776-4970 (outside USA)
Fax: 724-776-0790

Visit the SAE International Bookstore at books.sae.org

Publisher
Sherry Dickinson Nigam

Product Manager
Amanda Zeidan

Production and Manufacturing Associate
Michelle Silberman

Executive Summary

- The weight of a car is the best single measure of its environmental impact per unit of distance driven.

- Eighty-three percent of regulated pollutants that remain problematic are strongly linked to car weight: the *Molden-Leach Conjecture*.

- Consumers can use this as a simple way to choose the greenest vehicle.

- The total environmental impact of using a car can be approximated well by multiplying the vehicle weight by the distance driven.

- Governments can use this value to tax the environmental impact of car driving in a simple, fair, defensible and economical way.

- It can replace existing vehicle taxation, making the system less confusing, bureaucratic, contradictory and arbitrary.

- It can apply to all types of car, whether internal combustion engine or electric.

- It would be easy to implement as vehicle weight is public and information on distance driven could be collected annually through existing channels.

- Car weight and distance driven are values that are hard to falsify, making a future emissions scandal less likely.

- The new weight-based tax would start at 2 UK pence or 3 US cents per mile, or 1 Euro cents per km, in the first year for an average car, rising progressively to 8 UK pence or 11 US cents per mile, or 5 Euro cents per km, in proportion to the conversion of the cars on the road away from fossil fuels.

- Existing duty on *fossil* fuels would be maintained as a tax, and the level increased over time, ending at a level that would fully reflect the associated emissions.

- The combination of these elements would be technology and revenue neutral.

- This would incentivise lighter vehicles and less driving.

- Under the proposed system, taking the example of the UK, if an average car is 150 kg lighter, or does 1000 fewer kilometers per year, the owner would pay £100 less annually.

Contents

CHAPTER 16

Where Now?   247

Tax Base   248

Advantages   249

Summary   254

Foreword

We live in a complex world, and it is a paradox that the instant access to knowledge that the internet has given us has made it sometimes more difficult to form an opinion. The plethora of misinformation and cognitive bias that exists in so many areas of this wonderful resource requires us to examine the motives and the logic of the publisher more than ever. Equally, there is often a tendency to focus on a particular magic bullet to solve a problem.

This is probably seen no better than in the debate on climate change. The time for debate ended many years ago but still we have to endure emotive opinions and pseudo-science trying to persuade us that either there is no problem or that if there is then a singular solution will mitigate everything. In his previous book *Racing Toward Zero*, co-authored with Kelly Senecal, Felix showed the fallacy of such an approach.

In this book Felix and Nick Molden, having accepted that in the particular field of road vehicles the complexities are manifold, attempt to follow Einstein's maxim that "Everything should be made as simple as possible but no simpler." Of course, the danger of simplification is that of unforeseen consequences.

The authors are acutely aware of this danger as they try and find a simple solution to the vexed problem of replacing the duty currently applied to fossil fuels. In doing so they take the opportunity to rethink the direction of travel of the automotive industry while applying a fair tariff to the different powertrain and vehicle architectures that exist today. In doing so they certainly cannot be accused of over-simplification as they examine a multitude of factors beyond the obvious emission of carbon dioxide.

They question, quite rightly, whether a simple answer actually exists rather than just assuming that one does, and this rigor shows throughout the book. Ultimately, perhaps in an echo of Occam's razor, they find that not only does one factor correlate with most of the harmful aspects of vehicle construction and use, but that singular factor, in combination with a measure of usage, is perhaps the simplest measure to establish and police.

Pollution, be it carbon dioxide or the more complex and immediately harmful components knows, neither political nor geographical boundaries, and the authors ensure that they examine the consequences of their proposals in different territories, including North America and the Far East, rather than simply taking a Eurocentric view as so many do.

They develop the concept of the *Molden-Leach Conjecture* having examined the contradictory and illogical methodology used by governments worldwide to tax personal mobility. The conjecture provides a universal solution that is agnostic to the energy source, the territory the vehicle is operated in and, to a large extent, the manner in which it is operated. They propose a solution that does not just concentrate on any particular emission from the vehicle in use but looks at a full life cycle analysis of all factors which may have less than desirable factors and equates these to a simple, easily quantifiable factor.

Pat Symonds
Chief Technical Officer of Formula 1

About the Authors

Nick Molden is Founder and Chief Executive Officer of Emissions Analytics Ltd, a company he set up in 2011 to specialize in the measurement of real-world emissions from vehicles. The company pioneered the development of on-road test methodologies, including the launch of the EQUA Index in 2016. His primary interest is in bringing about effective solutions to environmental challenges and ensuring long-term sustainability of transportation. To that end, he has published many peer-reviewed academic papers, collaborates with universities around the world, presents regularly at industry conferences, and acts as an expert witness in court cases internationally.

Alongside his work at Emissions Analytics, Nick is a Founder and Director of the AIR Alliance, a not-for-profit organization that publishes free public ratings for the real-world emissions and interior air quality of vehicles. He is chairman of two standardisation groups within the Comité Européen de Normalisation: CEN Workshop 90 on urban tailpipe nitrogen oxide emissions, and CEN Workshop 103 on pollution ingress into vehicle interiors.

Nick holds an MA in Philosophy, Politics & Economics from the University of Oxford. He is also an Honorary Senior Research Fellow at Imperial College London.

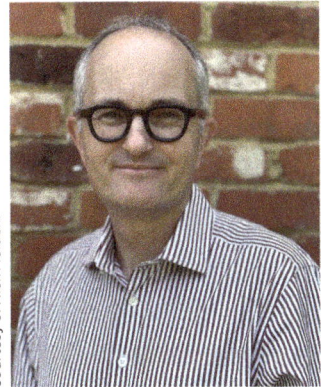

Felix Leach is an Associate Professor of Engineering Science at the University of Oxford, a post held jointly with that of Director of Studies, Fellow and Tutor in Engineering Science at Keble College. Felix has published over 70 peer-reviewed research papers - his research interests are in emissions and efficiency of thermal propulsion systems and air quality. He has had a particular focus on particulate emissions from gasoline direct injection engines and developing a fundamental understanding of NO_x emissions from diesel engines. Today, he works with zero-carbon fuels such as ammonia and hydrogen. He spent over a decade working in close collaboration with industry using world-leading measurement capabilities to help develop a clean engine for the 21st century.

Courtesy of Felix Leach.

In addition to his work on thermal propulsion systems, Felix has a significant engagement with public policy on emissions from vehicles and their interaction with air quality. He has run projects with Oxford Bus Company measuring emissions from in-service buses. He also runs projects, in collaboration with the Oxford councils, measuring and understanding indoor and outdoor air quality in Oxford and the impact of various vehicle restrictions on air quality and noise pollution.

Felix holds MEng and DPhil degrees in engineering science from the University of Oxford and is a Chartered Engineer and Member of the Institution of Mechanical Engineers, a Fellow of the Higher Education Academy, and a member of the Society of Automotive Engineers. Felix has been fortunate enough to be the recipient of many prizes, including for his first book *Racing Toward Zero*, co-authored with Kelly Senecal, published by SAE International.

Acknowledgements

We would like to thank Sherry Nigam and Amanda Zeidan of SAE International. Their help in guiding us through the book writing process and their words of encouragement along the way meant a lot to us.

Pat Symonds, Chief Technical Officer of Formula 1, kindly agreed to write the foreword to our book. Pat is a legend in motorsport, and we are grateful for his time and his words which set up our book so well.

Dr Richard Lofthouse has been a vital source of primary research and intellectual challenge.

We would also like to thank Dr Alan Kemp, Lord Heseltine, Steve Gooding, and Charles Leach for reviewing parts of our book and for their valuable feedback. Additionally, we would like to thank Dr Suzanne Bartington and the TRANSITION project as well as Gavin Turner for insightful discussions and creative contributions.

Nick is indebted to his family, Dr Helen, Harry and Arthur, for their vast wells of patience and unfiltered feedback. He would also like to thank all the staff at Emissions Analytics, particularly Eric Verdon-Roe for his extensive support and input.

Felix would like to thank his family, Eleanor and Nicholas especially, for their support. It is sincerely appreciated. Felix is also grateful for the many conversations with colleagues, graduate students, and friends that have informed so much of his writing in this book.

All power to those who actually want to solve the environment problems from cars.

Nick Molden
Felix Leach
August 2024

Introduction

Nick Molden

From setting up Emissions Analytics in 2011, and latterly cofounding the AIR Alliance, the road has been strewn with arguments with industry and lobby groups about some aspect of the company's measurements of real-world vehicle emissions. All fair game. But then an odd thing happened. Before, it was always a tussle of data, methodologies, test conditions, and vehicle selection, but conducted in a generally reasonable, measured, and rational way. Come battery electrification, this changed to arguments of fear and faith, conducted with accusations, personal slights, and sometimes threats. Governments responded by banning things—whether the internal combustion engine (ICE) as a technology, access to cities for certain vehicles, access to certain roads by any vehicle, and so on—often with debatable justifications. Bans may be necessary in exceptional circumstances, but, as an economist by training, my view is that this bar should be high, and more economically and socially efficient solutions are likely to exist. This book was born out of trying to understand how this fraught debate has developed, and what to do next.

When people become passionate, you know there is something important going on. The topic of climate change was already stirring strong emotions even before Dieselgate, but afterward, the debate moved rapidly up the gears. The passion was on many sides. There were, of course, those against the use of fossil fuel as the destroyer of the planet; those defending fossil fuels as the basis of our civilization; and the anti-car groups, the degrowth evangelists, and climate change deniers. Right now, we are engaged in a slugfest between lobby groups, leading to good policy being drowned out. Emissions Analytics was established to generate independent, real-world data to elucidate what really goes on, and I am personally motivated by finding the best and most efficient solution to this global challenge. We sit in the middle and try and work through the hype to find out the reality of emissions from transportation. This book presents a concept arising from that philosophy, which could become a key ingredient in the solution.

How could a debate on decarbonization—a fundamentally scientific one—become so unscientific? No amount of research or data is allowed to touch the totem of battery electrification. The more data and analysis that is thrown at it, the more intemperate the debate becomes. It is confusion that allows these unscientific positions to prosper, and in confusion malevolence can hide. Buying cars used to be a simple question of power and aesthetics. I will have a blue one with a big engine, please. Then came the push for diesel in the 1990s: go diesel and go green. "Clean diesel," in American terms. Then came the Dieselgate scandal and the complexity of vehicle emissions emerged, spiced with allegations of cheating and a loss of trust. Vehicle buying suddenly became complicated and laced with betrayal.

In the aftermath of Dieselgate, we did not know what the right thing to buy was to be environmentally conscious—except it could not be a diesel. Maybe it should be a hybrid. Enter the fairy godmother of electrification who could, with a swish of the wand, cleanse the past evils of internal combustion and point the way to a green future. A simple, compelling proposition. Enough people embraced this because they *wanted* these vehicles to be green. No evidence could be allowed to stand in the way. A simple solution to a complex problem. Deep-pocketed environmentalists found an enthusiastic constituency of disciples who would spread the word around the world. Repeat: Electric vehicles are zero emission. Keep it simple; maintain faith. Perpetuate the confusion as to the best solution, so most people have little option but to cleave to the simple battery panacea.

Many are frustrated at this turn of events, not least knowledgeable engineers with a strong grasp of the relative merits of different powertrains. However, as a class, these are often the very people implicated—directly or indirectly—with orchestrating Dieselgate in the first place. Their reaction has often been to dig deeper, offering up evidence to the benefits of internal combustion and demonstrating how clean the latest such vehicles are. None of this has achieved any cut-through. Why? Because all such arguments are complex. Complex as to the actual engineering solution—more aftertreatment and clever engineering. But also complex is how to explain why an internal combustion engine vehicle (ICEV) of 2023, with its aftertreatment system, is so much cleaner than the one of 2013, despite their both being ICEVs. Too complex. Too much inherent skepticism.

The question, therefore, became, whether there was a way to reset the understanding, expectations, and beliefs of consumers, commentators, and governments. Was there a simple truth underlying all the noise that would encapsulate, in a simple way, the new landscape of light-duty mobility, which is inherently a mix of electrified and analogue vehicles? Could Occam's razor be wielded to fight back for truth and balance in emissions reduction in transportation?

Success in this mission would go much further than revealing an underlying truth and establishing some interesting academic facts. It would go to the heart of the basic question of environmental intervention: Do you ban bad things, or do you put a price on them? This might sound like a relatively fine and arcane distinction, but it is in fact one of fundamental importance. Banning things requires a knowledgeable, near-omnipotent, and beneficent central actor sitting in judgment and deciding what to prohibit. Pricing pollution according to its damage leaves the difficult decisions of who does what to the power of individual decision-making—and raises revenue in the process. It leaves choice to free people using the "free market" of how best to abate pollution—those who are most willing and able to reduce their pollution do so, while those least willing or able do not have to, but must pay compensation to society.

This latter approach is the "polluter pays principle." It is an extremely powerful approach that in theory can lead to the optimal allocation of resources and behavior, thereby maximizing the welfare of society. Further, in theory, the pricing formula could be infinitely complex, but, in practice, it must be simple because governments look to implement taxes that are simple and high-yielding. Just as important, it is highly desirable in a tax that consumers can work out, in advance of their decision, the tax they would have to pay. This requires simplicity.

Together, all these points point toward a big prize. If there were a simple way to characterize the holistic environmental effects of cars, it would enable a price to be put on pollution and harness individuals' optimizing behavior to start mending the environment. Consumers would start being able to understand more instinctively the right thing to do. They would no longer have to grab simplistic memes about electric vehicles, and could make more sophisticated decisions. Governments would be able to create optimal tax structures that could help plug fiscal deficits and offset declining vehicle taxation. This would stop the rent-seeking behavior of well-funded lobbyists, using their deep pockets to influence complex rules, bans, and mandates to their private advantage.

Climate change is a big problem. Big problems need powerful tools. The market and free choice are those powerful tools, and they empower people, not lobbyists. This book, therefore, searches for the simple underlying truth in vehicle emissions that could unlock this process and end the current damaging environmental orthodoxy.

Felix Leach

As I sit writing this, we have just experienced the ninth month in a row of the hottest monthly average temperature recorded. My usual route to work has been impassably flooded for two of the last four weeks. Yes, the El Niño weather event is clearly playing a role in this, and, yes, weather is not the same thing as the climate. However, there is an overwhelming amount of evidence that climate change is here, now. It is not some random future phenomenon. That climate change is caused predominantly by human-generated carbon dioxide emissions into the atmosphere is also not in doubt.

For a variety of reasons, we are not changing our levels of emissions of carbon dioxide into the atmosphere anywhere near as fast as we can. Some of these reasons are good—immediately stopping burning things (the major source of global carbon dioxide emissions) would bring about a large amount of human suffering. Some are about technology—we do not yet have the level of cost-effective technology simply to make a switch and retain the same way of life. Some are about priorities—if your country is at war, should you be focusing your energies on carbon dioxide emissions reduction? Some are structural—countries cannot agree on who should reduce the most (my country emits less than 1%, your country over there emits more—but with 195 recognized countries, on average they all emit less than 1%) or who should pay for it, and within countries many people think it is someone else's job to sort out. And finally, some of these reasons are bad—a minority of people simply do not believe what I wrote in the first paragraph.

Carbon dioxide emissions come from a wide variety of sources, but just under 75% come from energy use, 18% from agriculture, 5% from industry outside of energy use (primarily cement), and 3% from waste disposal. Each of these sectors can then be broken down further. Break them down small enough and no part of the pie is more than 1%. This is a global, multisectoral problem that requires a concerted effort from a huge array of people in each and every sector.

This brings us to transport. I have worked for my entire career in reducing emissions and improving efficiency in the transport sector. This has taken a variety of twists and turns along the way, from looking at fuels to combustion system geometry. Today, that work still has many facets, but now includes new zero-carbon fuels such as ammonia—useful for decarbonizing large ships and for energy storage. In addition, I have branched out and also look at the problem of what happens to these emissions once they enter our environment: air quality. This has involved looking at what happens to air quality when traffic interventions are made such as a zero-emissions zone or low-traffic neighborhood, like we have where I live in Oxford. We have managed to link these air quality levels back to human health effects such as asthma—to my mind completing the chain from fuel going into the engine.

This is not to give you a CV, but to say I like to think I know what I am talking about in this area, and yet… When it comes to buying a car, which I have done in the past year, my decision was not easy. I am fortunate that I do more than 90% of my day-to-day traveling by bicycle—a benefit of living in Oxford—but that is not the case for the rest of my family and we do need a car. When a new arrival joined us, our Skoda Citigo no longer suited our purposes (to begin with, the child seat did not even fit in the car), so we needed to upgrade. What to choose? As you will have seen from this introduction already, climate change and the environment are at the forefront of my mind, so I wanted to make the best choice.

However, I am acutely aware that the environmental impact of a vehicle is not simply its tailpipe emissions, nor even its life cycle carbon dioxide emissions. I have also known about non-exhaust emissions (such as those from tires), and have been giving talks on them since 2017, there are also safety and noise concerns as well. I could buy a diesel—they are cheap in the UK at the moment thanks to low emission zones (in reality simply low nitrogen oxides (NO_x) zones) significantly penalizing them. To this day (and I reach for my flak jacket as I make this revelation), the lowest carbon dioxide-emitting vehicle I have ever owned was a diesel—and I got rid of that shortly after the Dieselgate scandal—yes, partly, out of shame (perhaps I should have more courage) and partly over the excess nitrogen oxide emissions. However, despite their arguably being the cleanest ICEVs on the road today (including under tests in real conditions), given the public perception and direction of the market, that did not feel right. A battery electric vehicle (BEV) or hybrid was appealing in many ways, but then I saw the cost. We simply could not afford either within our price range. That left a gasoline vehicle—and that is what we got, just like a lot of other people in the UK at the moment. Gasoline vehicles remain the top-selling category by far (both in the secondhand and new markets). Was that the right choice? I think so, but I really do not know, but also, I could not invest any more time thinking about it—we needed a new car!

And here is the thing, I did not *know* whether that was the right choice—and I know a lot about this subject. I was incapable of balancing all the nuanced environmental issues—carbon dioxide (both at the tailpipe and life cycle), nitrogen oxides, particulates, carbon monoxide, hydrocarbons including volatile organic compounds (VOCs), tire emissions, brake emissions, road wear, safety, noise, and impact on infrastructure. Perhaps I am just bad at my job, perhaps too much knowledge is a hindrance, or perhaps there is so much complexity in this topic that we need to rethink how we approach these decisions.

Consumers are faced with a lot of choice in this space. There are vehicles of many shapes and sizes—although fewer these days of the smaller sizes. How should a consumer who *wants* to do the right thing approach the choice? Information and incentives whether from the manufacturer or government are very complex and provide a cloudy picture. Is there a better way of easily assessing the environmental impact of a vehicle to help a consumer make this choice?

That is the starting point of this book—can we apply simplicity to reduce a really complex and uncertain choice to a straightforward one? Several things flow from this, of course. If it becomes easier to ascertain the environmental impact of vehicles, can we use this to influence people's habits and make it easier or even incentivize them to do the right thing in this space?

We also live in a world where the personal mobility space is rapidly changing. For over a hundred years, almost every car had an ICE. We also knew that most of the environmental impact of an ICEV (for most of those hundred years anyway) came from its tailpipe. Today that is far from the case. BEVs are making substantial inroads into the market and may eventually replace ICEs. Hybrids are well ahead of BEVs. Pure ICEV sales are softening.

This has consequences. First, we now need to compare a variety of different vehicle powertrain types—adding to our complexity and making it harder for consumers (and me) to do the right thing. Second, our whole regulatory—and yes, taxation—system is built on the tailpipe, directly or indirectly (we tax fuel which produces gas that exits at the tailpipe). What do we do when vehicles do not have tailpipes anymore?

I have known and had the pleasure of working with Nick for a long time. The work he has done at Emissions Analytics and the AIR Alliance is really first-class and impactful. He and I come at these questions in different ways, and I am very grateful—in particular—for his economics background—which I have none of! This book brings together our different perspectives to have a real go at solving these tricky problems. How can we decarbonize cars as quickly as possible? What do we do as the mobility space rapidly changes? How can we help consumers do the right thing? That is what this book is about—and together we propose the *Molden-Leach Conjecture* to answer them.

We also want to hear from you. If you meet us, let us know what you think. E-mail us. Contact us on social media. If you like what we have to say, advocate for it. We think it is a good idea and will try and persuade others that it is too—to change how we buy and drive our cars. Join us!

1

Auto-besity

IN THIS CHAPTER

- Car buyers want to do the right thing environmentally, but labelling information is too complex.
- Larger, heavier vehicles are increasingly common on our roads.
- Intuitively, heavier cars are bad for the environment.

This book is based on a premise and a question.

The premise is that, when buying a car, most people want to do the right thing environmentally, but the information and choices are now too complex for any normal consumer to really understand. The question is whether there is a simpler, more effective way to point the car buyer in the right environmental direction while enabling governments to tax and subsidize the right things.

The difficulty arises because environmental friendliness is ill-defined, multifaceted, and, to a degree, subjective. Some people glue themselves to roads to say we must "just stop oil" or "stop burning stuff." Others care more about water pollution or biodiversity. With a concept of environmental friendliness so broad and diverse, it is likely to end up meaning nothing rather than everything. Governments and agencies bring out more and more information under the umbrella of "consumer labeling," but little does any good, because it is either demonstrably inaccurate or overly complicated, so it leads to decision-making paralysis [1.1]. For example, despite the advent of food labeling, whether it be lists of ingredients or traffic light ratings for sugar and fat, obesity continues to increase in many countries.

Take the decades old European saga of labeling cars for their carbon dioxide emissions. To this day, the consumer label—meaning the values quoted in brochures and in advertising—underestimates the real emissions in normal driving [1.2, 1.3]. Starting with good intentions in the 1970s, a test was defined that reflected car performance then, but it was not updated until 2018 [1.4]. During that time, as the dynamic capabilities of cars improved, the test became less and less representative, and the gap between the emissions reported based on this test and real-world driving progressively grew [1.5]. By the early 2000s, the gap averaged 10%. As government policy then started to target reductions in carbon dioxide emissions, to slow climate change, carmakers needed the most flattering (lowest) figures and set about exploiting loopholes and gray areas in the testing legislation. The gap grew across the next decade and a half, exceeding 30% on average by the mid-2010s. Under pressure from consumers and politicians, authorities in Europe introduced an improved and updated test in 2018, but still left a gap of approximately 10% to real-world values [1.6]. This contrasted with the situation in the United States (US) at the same time, where the regulations ensured that official and real-world values were kept closely in line with the Monroney sticker consumer label [1.7].

So here we have a case of a relatively simple label in Europe that is needlessly inaccurate. If the Americans can make the system work, so can Europe. Yet, both systems are now being challenged by the transition to electric vehicles. With the ICEVs we are familiar with, a label telling you the emissions on a per-kilometer or per-mile basis made sense, as most of the emissions are caused by burning fuel to power the car. Not so with electrification, as most carbon dioxide emissions are created in the manufacture of the vehicle, notably for the battery [1.8]. Remarkably, whatever the level of these "embedded" emissions in the electric car, under both the European and US labeling systems they score a big fat zero. Big fat zero is apparently a good thing: zero emissions. Consumers are therefore told by the authorities (and manufacturers—Figure 1.1) how good these vehicles are, despite the zero label being demonstrably false. This, the official advice, is both oversimplified and wrong.

FIGURE 1.1   Zero emission—it says so right there.

fuadstephan/Shutterstock.com.

But herein lies the challenge. Should authorities update the label to include all emissions sources, such as battery mining, refining, assembly, and transportation? What happens if that makes the label very complicated, such that consumers get confused and switch off? In this book, we seek to find that sweet spot between something that is broadly faithful to the realities of vehicle emissions while retaining simplicity in usage. This is no small feat.

The challenges around carbon dioxide emissions labeling just set out reflect only one pollutant of many that arises from vehicles. Some come from the tailpipe, some from other parts of the vehicle, and some from its manufacture and end-of-life disposal [1.9]. Some—like tires—emit particles when in use, and also give off potentially toxic fumes even when not in use. A label capturing all the main pollutants would create one so extensive and baffling as to be useless, particularly when there are trade-offs between pollutants—they do not necessarily reduce together. On the other hand, if the focus were put on too few, pollutants and critical environmental effects would be missed.

The flip side of the challenges facing consumers, as they switch from traditional cars to electric vehicles, is the challenge governments face raising taxation and recouping the environmental costs of cars. Governments must raise taxes to pay for public services, and taxing vehicle emissions is fertile ground, because it can be justified on the grounds of reflecting the environmental damage caused. Without such taxes, cars would be allowed to pollute common goods like air and water, damage human health and worsen the climate for free. We all need to pay to get our garbage collected from our homes, we ought to pay similarly to dump our emissions trash. Compensating the rest of us for this "externality" in economics terms, is the simple correction of an historic market failure [1.10]. Placing a tax on driving in some way can ensure the driver pays for the externality and, therefore, is incentivized to drive the optimal amount. For such taxes to work well, they need to be simple to operate, hard to cheat, and yield sufficient revenue to cover the externality. The ideal environmental label should help achieve all three elements.

Environmental Labeling

While environmental labeling may seem arcane and technocratic, its power should not be underestimated. That BEVs currently come with a zero-emission label is strongly driving the market, despite the label not being accurate. This costs both the planet and the exchequer. It is not so much that the zero-emission label directly motivates the consumer to buy an electric vehicle when otherwise he or she would go for gasoline (petrol). Rather, the power comes from how it interacts with the law and taxation policy. In Europe and some states in the US, there are targets that manufacturers must meet for selling "zero-emission vehicles" (ZEVs) (although the exact definition varies slightly from place to place) [1.11]. Miss the target, and the carmaker has to pay substantial fines. As a result, they will push and, if necessary, discount the price of electric vehicles in the showroom—even if the consumer does not enjoy a benefit in making the switch. In some countries, there are retail incentives to buy BEVs—for example, $7500 off the sticker price— either as a cash subsidy or a tax credit [1.12]. Less overt, yet possibly even more powerful, are incentives through the "company car" system, through which employees are given substantial tax benefits on a car provided to them by their company. This has been a signifi- cant driver in the early uptake of BEVs in the United Kingdom (UK), for example [1.13].

The combination of these strong incentives, based on a funda- mentally flawed label, can only lead to far-from-optimal environ- mental and fiscal outcomes. The evidence is already out there on the road. What we observe on the highways in both Europe and the US today is ever bigger vehicles. Evolving consumer tastes have rein- forced on-going ill-conceived labeling and taxation. Large sports utility vehicles (SUVs) are bought with carbon dioxide labels in Europe that systemically underestimate these emissions [1.4]. Wrinkles in how the average emissions of manufacturers are calcu- lated—which are relevant to government-set targets and fines— incentivize larger vehicles. BEVs, typically up to 40% bigger and heavier than equivalent ICEVs [1.14], have reached almost 20% of worldwide sales in a relatively short time [1.15]. We are living in an age of auto-besity, without the slightest prospect of being put on anything resembling a diet.

We should note that throughout this book we are going to use the terms mass and weight interchangeably. We know that they are not the same and that weight is a force, however, in line with common usage, and to avoid some clunky sentences, we use the terms interchangeably.

Weight Gain

No matter how much drivers like them, it is intuitive that we should be skeptical that making bigger cars will save the environment and our quality of life. Consumer tastes have shifted markedly to the SUV because of their high driving position, ease of entry, and perceived greater safety. For all of that, these vehicles have more energy to dissipate in a crash, wear the road infrastructure more quickly, and generally require more materials to build. The physics is unambiguous: it takes more energy to move a heavier thing around than a lighter thing, and energy use is well correlated with emissions. Perhaps a large battery replacing the ICE, while much heavier, will prove to be the key to emissions-free motoring. But as this conclusion is counterintuitive, the evidence must be strong. In this, there is a lesson from recent history. When Dieselgate—the Volkswagen emissions "cheating" scandal in 2015—broke, it turned out that most manufacturers in Europe had been manipulating their nitrogen oxides (NO_x) emissions results [1.5, 1.16]. Much of the problem was with smaller diesel cars, which did not have the space to deploy the cleanup systems in the exhaust, even though the technology existed [1.17], and their profit margin could not justify the expense. So, circumventing the rules became the preference. The exception was larger diesel vehicles, which had more space and profit margin, and so this was where the proper cleanup technology was mainly deployed [1.16]. In this case, bigger aligned with being more environmental. But perhaps this is the exception, rather than the rule.

We have not always been suffering from auto-besity, and not all countries do even today. Rapid expansion of personal mobility came to Europe after World War II, where small frugal cars were built out of the available raw materials. Europe exported these cars around the world in large numbers, except to the US, which had already tended toward larger vehicles because of cheap gasoline. Japan, in contrast, an importer of oil, tended towards the smaller, frugal "kei cars," followed 50 years later by hybrid vehicles made famous by the Toyota Prius in the late 1990s. Nevertheless, by the turn of the millennium, Europe and Japan were following the US example as consumer preferences, understated emissions figures, and other regulatory biases came to the fore. Unsurprisingly, the misconceived current system and incentives in favor of SUVs and BEVs accelerated the trend in recent years [1.18].

Having highlighted the continuing tendency of cars to gain weight, and suggested the environmental consequences, we now need to set the phenomenon in context by reviewing the various impacts of cars on the environment—not just tailpipe emissions, but other emissions, for example from tires, together with phenomena such as safety and noise, and even infrastructure impacts. We will consider the degree to which the weight of vehicles influences environmental outcomes. To those of a more populist turn of mind, we will not, however, suggest that battery electric cars are exclusively responsible for making car parks spontaneously collapse or bringing about the end of Western civilization.

In 2022, for the first time in 14 years, the Volkswagen Golf—a midsized hatchback car—was dethroned as the highest-selling car in Europe, by the Peugeot 208—a similarly shaped vehicle [1.19]. In 2023, the Tesla Model Y, a BEV, took top spot [1.19, 1.20]. The typical mass of a current Golf is approximately 1500 kg (3300 lb)—this is known as the "kerb weight," which is the weight of the vehicle including all accessories and some fuel in the tank, but not passengers or added payload [1.21]. For historical reference, the original Mark 1 Golf launched in 1974 weighed approximately 750 kg (1650 lb) [1.22]. The Tesla Model Y is approximately 2000 kg (4400 lb) [1.21]. Tesla's success owed a great deal to Europe's policy of pushing electric vehicles. One unintended by-product has been to increase further the weight of the best-selling car by one-third. The US was not so very different. From 2002 to 2020, the best-selling car in the US was the Toyota Camry, a 1700 kg (3750 lb) midsized sedan [1.23]. In 2023, while the car pack was led by the Toyota RAV4 SUV, the Tesla Model Y came a close second, its weight exceeding the Toyota's by 300 kg (660 lb), not far short of 20% [1.24]. Next, consider pickup trucks, an icon of US motoring, and a market led for 40 years by the Ford F150. Its combustion engine version weighs approximately 2600 kg (5700 lb), while its new electric version—the F150 Lightning—adds an additional 300 kg to reach 2900 kg (6400 lb) [1.21].

There is nothing exceptional about the examples given above. They are representative of the wider market. In the US, according to the US Environmental Protection Agency (EPA), average vehicle weight has grown by approximately 25% since 1980 and continues on an upward trajectory, as shown in Figure 1.2 [1.25]. Unsurprisingly, the footprint of vehicles—the silhouette on the ground—has also gone up by approximately 5% since 2008. This indicates an expanding girth (and length) as well as weight. But note from Figure 1.2, the trend in the 1970s. In the seven years following the oil crisis of 1973, vehicles became dramatically lighter and more fuel efficient, in part by sacrificing power. The average weight of a US vehicle fell by more than 20%, power was reduced by 25%, and fuel economy surged by approximately 70%, reducing the associated carbon dioxide emissions by around 40%. As a result, not for environmental reasons, but from the impetus by consumers to save money and by the government to protect its geopolitical interests, a key part of the environmental impact of vehicles was reduced by two-fifths as a result of a one-fifth reduction in vehicle weight. The 1980s saw a reversal of this trend as fuel prices fell and regulations started to encourage larger vehicles, especially pick-up trucks.

FIGURE 1.2 Average vehicle weights, footprints, fuel economy, and power 1978–present. (Data, US EPA) [1.25].

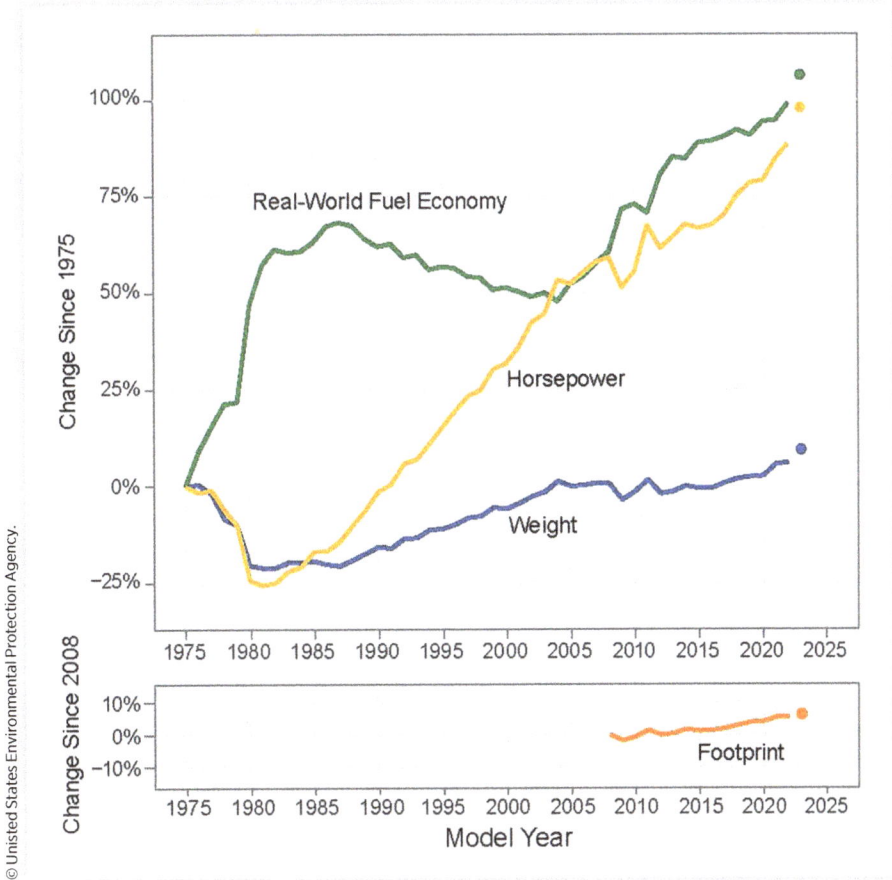

Data from the European Union (EU) points to a similar weight trend over the last two decades. In 2001, the average weight of an EU vehicle was approximately 1275 kg, rising to 1393 kg in 2011 and 1457 kg in 2020, according to the International Council on Clean Transportation, as shown in Figure 1.3 [1.26]. This represents an increase of 14% over the period, exceeding even the increase in the US. The chart shows similar trends across the EU. After the financial crash of 2007–2008, a similar, although not as extreme, reaction to the 1970s oil crises was seen, with average vehicle weight falling in a search of fuel efficiency.

FIGURE 1.3 Average vehicle masses in the EU by country 2001–2021 [1.26].

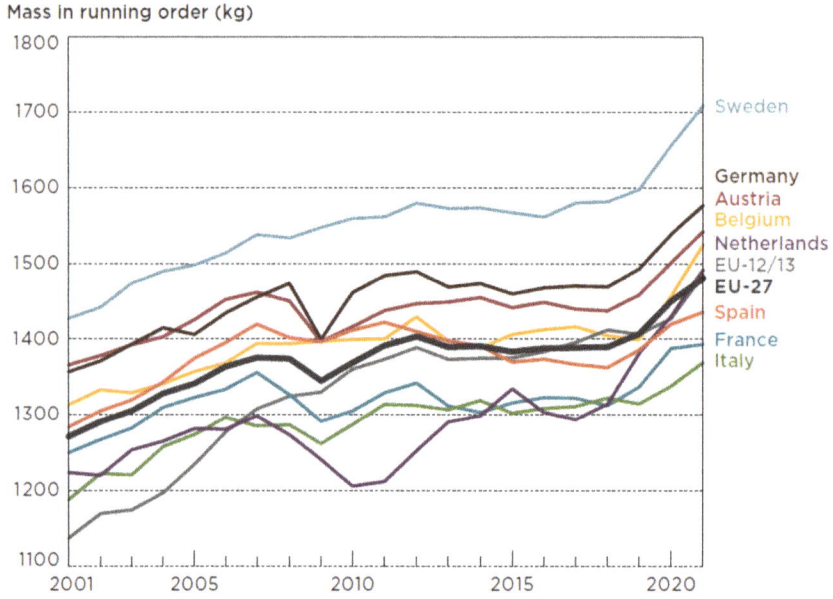

Figure 1.4 breaks down the same data by the main vehicle manufacturers and reveals a contrast between the premium German manufacturers and most mass-market companies. In an attempt to fulfill increasing consumer preferences for large, high-riding vehicles, and take a lead in electrification, BMW, Mercedes-Benz (part of Daimler AG), and Audi (part of Volkswagen AG) led rapid increases in vehicle mass from 2018. After a period of increases, French and Italian manufacturers saw average mass fall over the same period, reflecting in part a slower start to electrification.

Mass data for other countries are not as extensive and available as for the US and Europe, but the trend in China is worthy of note. Just as in its more developed peers, the last two decades in China have seen a marked shift from small city cars to SUVs. As a result, the average weight of cars increased by 23% to 1476 kg (3250 lb) between 2005 and 2019, bringing China to near the global average [1.27].

FIGURE 1.4 Average vehicle masses in the EU by manufacturer 2001–2021 [1.26].

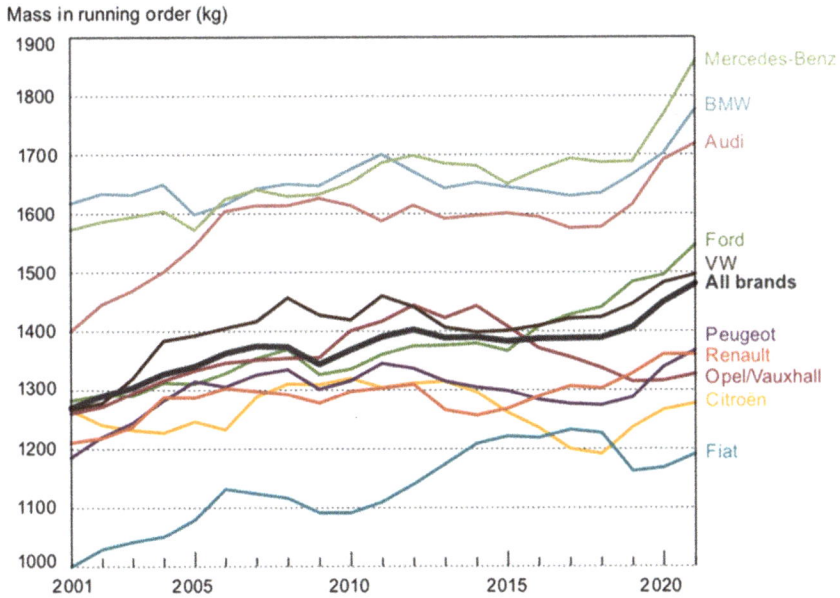

Mass in running order (kg)

Mercedes-Benz
BMW
Audi
Ford
VW
All brands
Peugeot
Renault
Opel/Vauxhall
Citroën
Fiat

The term "auto-besity" suggests that bigger is not better, but is there an intuitive basis for this? This book will examine whether there is strong empirical support for the idea, but does it even pass muster on a quick view? We think it does. Fundamentally, the reason stems from the essential notion that a vehicle must carry its own weight. In contrast, a power station does not. The mass of the metals and bricks in the edifice of a power station has no significant bearing, in themselves, on how clean the electricity it is producing is. No input fuel is being used to support the structure of the building. In contrast, a car—like a ship or an aircraft or indeed a human—must carry both its own weight and the fuel to power it. These two elements in fact reinforce each other: the heavier the craft, the more fuel must be carried on board to go the same distance as a lighter one.

Casting our minds back to school physics lessons, we know that the term "work" is not just about laboring away, but it is a strict concept about energy transfer. Energy is transferred to an object by the application of a force through a distance the object is moved. Force times distance. (Or, more accurately, force times displacement, but that is an unnecessary detail for here.) To drive from Paris to Berlin, other things being equal, a Bentley will need more work doing than a Toyota Yaris, because of the weight differential. With the effect of having to carry your own fuel, the Bentley may even have to do disproportionately more work. Therefore, by this simple application of physics, we can see that it is a reasonable hypothesis to equate weight to work done. But why is work done bad for the environment? In a car, the work is done by the engine, and the engine needs fuel. In the case of a BEV, the work is done by the motors, and the motors need electricity. More work, more fuel or electricity. Liquid fuel and electricity are valuable as they are energy carriers, so more work means more energy is required.

Challenges

This is where the first complexity—of many—comes in. To do the work to move the vehicle, we need energy, which is stored in the gasoline, electricity, diesel, methane, or whatever fuel is being used. But the amount of fuel needed varies significantly between types of propulsion. A gasoline engine is only up to 35% efficient in extracting the embedded energy and turning it into useful work to move the vehicle along [1.28]. Much of the rest is lost as heat to the environment—witness the hot engine block and exhaust gases. This is a feature of such an ICEV, which is subject to complex but inescapable thermodynamic limits in the laws of science. We are told that BEVs are so much better at the efficiency game, turning approximately 90% of the energy in the onboard battery into motion [1.29]. This is true, but it is necessary to subtract up to approximately 30% for energy losses when charging up the battery in the first place, which again is lost mostly as heat [1.30]. Add to these losses the inefficiencies in the power stations creating the electricity to store in the battery, and the overall efficiency gap with combustion engines can almost disappear, as we will look at in more detail in a later chapter.

If less fuel is needed to move the vehicle, that should equate, at an intuitive level to fewer emissions. A second complexity comes with the extensive pollution cleanup technology that is deployed in the exhaust of a modern combustion engine. Various "catalytic converters" and filters (collectively referred to as aftertreatment) can neutralize much of the pollution created by the onboard combustion, to the extent that it disconnects the environmental effect from the energy used and work done [1.31]. These are just two of many complexities, and, therefore, while our intuitive link between weight and emissions is solid, we must acknowledge that it may not stand up to the real-world scrutiny that will follow in subsequent chapters.

Zero Emission

Electric vehicles are not just a complicating factor for our intuitive concept; they may offer a fundamental falsification of it. The official "zero-emission" label of BEVs points very directly to this: if they are zero-emission, then their added weight is *ipso facto* not an issue. It is increasingly understood, however, that these vehicles are not zero emission, not least because the mining and processing behind the batteries creates significant emissions. Nevertheless, so long as they provide significant reductions in carbon dioxide emissions over their lifecycle, this may be enough to invalidate our hypothesis. This is why this book will systematically consider not just carbon dioxide and not just the tailpipe pollutants currently regulated, but a wide range of unregulated pollutants and other effluvia. We go on the hunt for the real environmental downsides hidden behind comfortable zero-emission labels and received notions.

In this hunt, there is one further conceptual dimension to hold in mind: time. If we look at the trend in ICEVs, whether over the last decade or century, they have been getting heavier but cleaner [1.26]. Early ICEVs were extremely inefficient and spewed untreated exhaust gases into the air [1.28]. Catalytic converters started coming onto the market in the 1960s, which led to a substantial reduction in emissions, just as vehicles were starting to become bigger and heavier on average [1.31]. US tailpipe regulations since then have progressed from Tier 1, to Tier 2, and now Tier 3—ever-tightening limits on emissions [1.28]. In Europe, the "Euro" stages of emissions regulation started in 1992 and now are at "Euro 6," with "Euro 7" set to follow from 2025 onward. Is not it the case, therefore, that this perspective over time clearly invalidates the hypothesis that increasing weight correlates with more environmental impact, and points in fact to the opposite?

When we are talking about consumer labeling and vehicle taxation, we are concerned first and foremost with the optimal choice between new vehicles on sale today. What is the most appropriate car to buy from those currently being offered by manufacturers? For these purposes, we are not interested in consumers making a choice between buying a new Toyota Yaris or a classic 1960s Ford Mustang. That is not a choice a consumer would typically make. It is true that a consumer might weigh up buying a new Toyota Corolla against a two-year-old Honda Accord, but those two vehicles nevertheless are approximately the same age and generation of vehicle. In other words, we are looking for useful information that helps make the right environmental decision and set the right taxation among vehicle choices that are being made in practice today. That, in the long sweep of history, vehicles have become heavier but cleaner is interesting but irrelevant. We are concerned about whether, among those vehicle choices being made, greater vehicle weight correlates with more environmental impact. In fact, it is possible for there to be both a negative correlation between weight and emissions over time and a positive correlation between weight and emissions at the current instant in time. We are just interested in the latter. Put another way, it is well established and a positive trend that cars get cleaner over time, but what matters is to choose the least environmentally impactful vehicles *of those available on the market today.*

Summary

In conclusion, it is clear that vehicles are becoming heavier in most major car markets of the world. By extension, the trend is likely to follow in other markets, as many models are exported there. We have an intuitive concept that the fattening of cars is an environmental bad, as it leads to the use of more fuel to move the vehicle and the physical impact on its surroundings is greater. However, BEVs present an apparently problematic case, as they are heavier but use less energy, and sweep around our streets near-silently and with only the occasional tap on the brakes. We must examine this in detail in the coming chapters, but also question whether there was a weight-dependency even with ICEVs. By looking at each type of environmental impact in turn, we can test this hypothesis and see if there is a better candidate for that single piece of information to guide consumers and governments.

This all matters because if decisions are not based on good information, wrong choices are likely to be made. Consumers may have an illusion they are doing the right thing, but they often may not. Governments may think they are internalizing the environmental externality effectively using the current tax system—apply an appropriate "sin tax" in other words—but they could be causing even worse distortions. We are in search of something that is simple, effective, easy to implement, and hard to cheat, that is the best directional pointer for the right behavior in the complex and baffling world of environmental concerns. In doing so, by proposing something simple and broadly based, we aim to avoid creating a narrow solution to one problem and, in doing so, creating some new distortion elsewhere.

2

Thinking about Environmental Impact

IN THIS CHAPTER

- As it is hard to obtain sufficient information on environmental impact, we use rules of thumb, but these need to be as good as possible.

- Banning "bad things" sometimes works but often has significant downsides.

- Applying taxes to environmental externalities is a superior way to achieve rapid environmental change.

The environmental impacts we are considering in this book all stem from the consumption of goods and services: the motor car and supporting products and services. Environmental gain can be achieved by addressing the demand for these goods and services, the supply, or some combination of the two. In other words, policy or personal action can reduce the quantity consumed of goods and services that are "bad" for the environment by forcing or incentivizing consumers to want less, or by making it less profitable or limiting the amount that companies can sell. Before readdressing the specific question of vehicle emissions and taxation, we need to consider in a wider sense how individuals and companies think more generally about environmental impact in their decision-making.

Only in rare cases would a consumer—quite understandably—go and perform some primary research before making a purchase. It would be unusual to visit a factory, interrogate the company's annual report, review any published research, and so on. Even if this were practically possible, it would involve significant work, raising the effective "price" of the good, thereby reducing welfare. Considered another way, competitive markets rely upon getting as close to "perfect information" as possible on their preferences, product specifications, and market prices, and so any lack of information to the consumer on the environmental credentials of products would lead to a suboptimal market outcome and welfare. Put another way, we might get fewer products and higher prices than if we had better information. So, the absence of the necessary information, or the work required to plug the informational hole, is an undesirable situation for our happiness.

In reality, though, this is the natural situation. There is information about the environmental features of a product that the consumer cannot immediately access, at least not without some intervention from the government, authorities, or other players. The beauty of the market economy is that the price of a good is meant to be a singular expression, or abstraction, of the demand and supply considerations of all the buyers and sellers in the market. It is a hugely rich informational summary, which has been largely responsible for driving formidable increases in social welfare across the globe, particularly since the Industrial Revolution. However, this price "signal" can malfunction. If the visible price is not actually the true price, this beautiful mechanism for allocating resources breaks down. If a tire manufacturing facility dumps the waste from its production process in a nearby river, that is costless to the company but negatively affects surrounding towns and wider society through loss of amenities of health impacts. If a consumer throws a wet wipe—the moist towelettes beloved of new parents— down the toilet, the resulting fatberg that blocks the communal sewer has to be fixed at a cost to the taxpayer rather than the careless parents.

These are called environmental "externalities" by economists [2.1]. In other words, there are costs that the production or consumption of the good imposes on society, including the environment, that are not priced into the production cost or sale price. The price is then too low and, as a consequence, we tend to consume too much of the bad good. That is, in essence, what happens with cars and their pollution—we are often not paying for that pollution, so buy too many cars and drive too far. People of certain, often more anti-market, persuasions would see this "market failure" brought about by the externalities as a reason to abandon the market system and set about banning things. This is, however, a gross overreaction as it is possible to solve the externalities problem. It is possible to correct the failure by pricing in the externalities, often through some sort of tax [2.2]. To take a simple example, if, for every wet wipe sold, we knew that the cost of sewer clear-up and loss of amenity to users were equal to 5% of the price of wet wipe, a 5% "wet wipe tax" would fix the pricing signal such that consumers could decide on the right number of wet wipe packs to buy.

Such an approach would, in principle, be superior to banning wet wipes, in the way that plastic straws have been in many places. While a ban should sort out the wet-wipe-blocked sewers, the reduced happiness of parents having to use inferior and less convenient cleaning materials would be significant, although possibly be underrecorded and appreciated. Quite likely, alternatives to wet wipes would be developed and sold to circumvent the ban, which could be even worse for the environment and consumer utility. Witness the sad, limp cardboard straw lifted from the Coke can. While motivated by a desire to reduce single-use plastic waste, the often-plastic-coated straw may create an equivalent waste problem; the straw may have caused even more carbon dioxide emissions in its production than the plastic one it replaced, and that is before a value is put on the exasperation of the drinker. Thought of another way, a ban is effectively an infinite tax. The product is made of infinitely high price such that demand is zero. We know for certain that the sewer cleanup cost is not infinite and, therefore, it is quite logical that a ban must be a suboptimal solution. We shall return to the subject of banning the ICE, and its inherent inefficiencies, in later chapters.

Taxing Externalities, and Shorthands

If we can move past the lazy desire to ban bad things—which surely should be reserved only for the most egregious dangers—we can see that taxing externalities has many other advantages. Although not the case so often, externalities can also be positive and so can justify the application of subsidies—education and its spin-off benefits on civilized society are one of these [2.3]. An important further flexibility is that multiple taxes—or subsidies—can operate at the same time in the same market simultaneously to correct multiple externalities. This will be highly relevant with ICE motor vehicles, which emit both greenhouse gases and air quality pollutants, among other externalities. Different things, produced under different types of driving, can be corrected by different taxes operating in parallel.

While this solution sounds simple in principle, it can be complicated in practice. As mentioned at the start of this chapter, consumers will not typically perform primary research to quantify the environmental impact—the negative externality—and then adjust their quantity demanded accordingly. Too much effort. Too little easily available information. To resolve this, they look for two solutions. First, are there shorthands, shortcuts, or rules of thumb? Second, can an authority step in to provide the necessary information? While the second option sounds attractive for getting the best answer, it is still necessary not to burden the consumer with large amounts of information that he or she then must assess. As a result, authorities perform the research and then condense the results to simple information or ratings or consumer product labels. So, either way, we end up with shorthand summaries to make the consumption or, if it is on the supply side, for companies to make the production decision. This process of compressing information into these shorthands inevitably means that detail is lost, which is the price of the simplicity and resulting practicality and utility for consumers. If oversimplified too far, these ratings would not function as the basis for taxation, or other interventions, to correct the environmental externality. If not simplified enough, consumers would be baffled or at least put off the work to analyze and would then continue to make decisions without any relevant environmental information. Either outcome would be poor.

Therefore, simplification into a rating must be optimal in at least two ways. First, it must be in the Goldilocks zone of neither too condensed nor not condensed enough. Second, the information sacrificed in the simplification must be the least material to environmental impact. For example, carbon monoxide emissions from vehicle exhausts are now well controlled so as to be a minor contributor to air quality problems, and so could potentially be sacrificed for labeling simplicity, while still being officially regulated and controlled [2.4]. Much weight is put on authorities constructing ratings according to these principles. The Dieselgate emissions scandal in 2015 is a good example of when it goes wrong [2.5]. Air pollutant emissions from vehicles were classified under the "Euro" standards scheme, which developed—supposedly—ever-tighter limits over time, all summarized in one number. For example, when Dieselgate happened, all new cars had to be certified to "Euro 6." That label carried the weight of describing all air pollutant emissions. While most vehicles were fine for carbon monoxide, hydrocarbon, and particulate emissions, there was a large variability on nitrogen oxide emissions [2.6]. But as long as emissions were below a certain limit on the official test, a Euro 6 pass was given. The problems were twofold. First, by not breaking out a vitally important pollutant, nitrogen oxides, it was impossible to apply any form of tax or other corrective measure on it alone. Second, the system was structurally weak such that many diesel cars were in fact not meeting the limits at all, despite being given the "Euro 6" sticker. Therefore, the rating system was badly structured—too much information was lost—and it was badly policed so that even if nitrogen oxides had been split out, the information would have been misleading anyway!

More widely than the subject of vehicles, we create and rely on shortcuts around the environmental credentials of products to help catalyze decision-making. Plastic is bad. Beef is bad. Wind power is good. Bicycles are good. Abstention is good. Nuclear is … complicated. The National Westminster Bank in the UK introduced a feature in its mobile phone banking application in 2023 that analyzed and categorized your individual transactions and ascribed a carbon footprint to each [2.7]. It goes further and makes a recommendation as to how you can reduce your environmental impact. Use less plastic! France in 2022 banned the purchase of patio heaters on the grounds of the unnecessary and unwarranted carbon dioxide emissions in their operation [2.8]. Single-use plastics, including straws and plastic bags, have become subject to a variety of bans and taxes in many countries in recent years [2.9, 2.10]. Interestingly, some countries, such as the UK, have gone for taxation—typically it is 10 pence per bag—while others such as Italy and China have preferred bans [2.9]. As discussed previously, the former is more economically efficient than the latter. Strictly, this allows for allocative efficiency, where people consume up to the point that their marginal benefit is equal to the true marginal cost.

These ratings, initiatives, and recommendations all sound virtuous, but are they effective? For some, perhaps, the question of effectiveness does not arise. For those for whom virtue-signaling—indicating to others your moral superiority through your observable purchase decisions—is the priority, whether an environmental benefit is, in fact, achieved in the end is of very much secondary importance. For these people, it is about the perception of others. If single-use plastic is generally seen as environmentally bad—whether or not it is—that is sufficient to signal virtue. For a much wider group of people, while they may be time-pressed and making large numbers of decisions every hour, there is a hope and belief that their decisions are reducing environmental impact, even if they do not have time and access to information to verify. For these people, it is less about signaling to others, but more about belief and trust that the shorthands they are using do in fact correlate to a significant degree with the environmental benefit. This leads us to the question: Where do these shortcuts come from, especially where there is no obvious rating system provided by the government?

Shorthands in Other Sectors

Setting aside vehicles for now, shorthands can exist for everything we produce and consume. The sectors that are responsible for significant emissions include agriculture, construction, electricity production, media, and plastics. Between these sectors, including transportation, they account for a substantial majority of world-wide carbon dioxide emissions, together with a combination of local air pollutant emissions [2.11]. Considering agriculture first, we have already mentioned the shortcut that beef is "bad." Pork is not quite as bad, and chicken—if you have to eat meat—is better. Vegetarianism is even better, while you cannot beat the apparent moral superiority of veganism. Within the choice of meats, this classification is well founded, as beef is responsible for approximately 37 kg of carbon dioxide per 1000 kcal of meat produced. Pork is typically 85% less than this, and chicken is a further 25% less [2.12]. The shorthand for an environmentally conscious decision is the *type* of meat. This is generally well-founded in data, and the shorthand is attractively simple for consumers to base their decisions on. Such an approach could even lead to a simple taxation system where beef attracted a tax higher than one imposed on pork and chicken.

It gets more complicated when this shorthand is extended to vegetarianism and veganism, or other dietary regimes. The problem is that the word "vegetarianism" is not defined with a sufficient and relevant level of accuracy. Some vegan diets struggle to deliver the necessary levels of daily protein for good human health. So, while on the surface vegan diets may involve fewer emissions in their production and transport, the consumer may be sacrificing health and utility in such a way that the scenario is not comparable. A vegan diet delivering equivalent protein and calories as a meat-based one may not be as environmentally beneficial as expected. In short, using a shorthand based on some poorly defined terms relative to a complex combination of foodstuff and behaviors is a poor basis for decision-making. Choosing vegetarianism over meat-eating is certainly a simple rule of thumb, but the terms may not correlate as well as expected with the underlying environmental benefits sought. It is difficult to imagine how a tax or subsidy structure could be based on such dietary concepts. A better approach may be to consider each foodstuff individually, but that risks losing the necessary simplicity to be actionable.

An alternative way to approach agriculture is from the production side. Three big contributors toward emissions and wider pollution are the use of fertilizers and pesticides to maximize yield, and of fuel to facilitate mechanization and maximum productivity.

Fertilizers are a major source of global carbon dioxide emissions, accounting for approximately 2.1% in total, and more than 10% of total agricultural carbon dioxide emissions [2.13]. The proportion is high because of the energy-intensive Haber–Bosch process typically used to produce the fertilizer [2.14]. While alternatives are being developed, those do not yet exist at scale. Currently, the best way to reduce this is to reduce the amount required in the first place by more careful application. Pesticides produce less carbon dioxide per unit; although the active chemicals are formulated in the laboratory before being mixed into the carrier, the energy required is less [2.15]. For pesticides, in particular, the environmental effects go beyond the carbon dioxide emissions, as was seen with the believed impact of neonicotinoids on bee populations in many European countries [2.16, 2.17]. The final significant component in agriculture is the fuel required to power the mechanization—to a large extent diesel. Arable farming, in particular, requires cultivation of the ground and then harvesting using diesel-powered machinery to achieve optimum yields. All parts of agriculture use fuel to store and transport the end products [2.18]. Whether it is fertilizer, pesticide, or fuel, there is a simple proxy to allow us to estimate the environmental impact of each: it is the *amount of stuff* used. This may sound trivial, and in a sense it is. It is blessed with simplicity, unlike other shorthands such as "vegetarianism," which give little information on their own. The amount of fertilizer applied is the key piece of the information; not so much where it is applied or for what purpose, just the amount used.

The construction sector shows many similarities to agriculture in terms of these shorthands. From the supply side, some of the big emissions contributors to construction, in the round, are concrete, steel, and fuel. Concrete contributes to approximately 8% of total global carbon dioxide emissions [2.19], followed by steel at approximately 7% [2.20]. There are, of course, many other metals that together contribute significant emissions too, including iron, copper, and aluminum, as they all require mining, refining, and then forming the end product. Construction projects are seen first and foremost through the eyes of the quantity surveyor—considering the price and, crucially, the quantity of each input material. The *amount of stuff* used is, as a result, a fairly direct way to estimate the material and energy and therefore the environmental impact.

But this is not the approach used by consumers. Take your home. How do we think about its environmental impact? In Europe, the Energy Performance Certificate is perhaps the most prominent tool [2.21]. The dwelling is rated for its energy efficiency on a scale from A to G, and this result is presented on an official label with great authority. However, it gives a fragment of the real picture. While how badly your house "leaks" heat undoubtedly affects the amount of gas, oil, or electricity used to keep it warm, other factors are ignored. Most importantly, the absolute size of the house matters. An A-rated palace would rightly be judged worse for the environment than a G-rated studio apartment, as the former would use a larger absolute amount of fuel for heating. Moreover, the palace would have higher embedded emissions from its construction, although these emissions may be small relative to heating costs as they are spread over the lifetime of the building. Another shorthand that might be used is the presence of certain technology in the house—an air-source heat pump for heating or a solar panel for electricity generation. However, such simplistic classifications would ignore the renewable energy storage capacity, the quality of installation, and so on. As the flip side of the quantity surveyor approach on the supply side, it is perhaps the floor area of the property that should be the best proxy for environmental impact, which would give a dramatically different answer to the Energy Performance Certificate.

Fossil fuels are not just used to heat buildings and power homes, but also produce the plastics we enjoy in modern society. And plastics are complicated. Complicated because they come in so many different forms and are used in many different ways. As many plastic items are small, they provide an end-of-life collection problem. Whereas a demolished house or scrapped car is in a local-ized and predictable place, a plastic bottle may be discarded anywhere, in an unpredictable fashion. Through its life, plastic creates emissions of many sorts. First, carbon dioxide and other pollutants are released during its original fabrication and distribution. Although a smaller part of the total effect, plastic products can release toxic chemicals during their use. A group of chemicals called phthalates are used to make plastic products more durable and flexible, but if they are ingested directly or indirectly via food and drink, they can have negative reproductive and neurological effects on humans [2.22]. Once tossed away, either to a grassy verge or landfill via the refuse system, the plastic gradually breaks down into microplastics, which can cause significant problems for marine creatures and the wider marine environment [2.23]. The microplastics can be ingested by and accumulate inside fish. While there, or in the water, a wider range of organic chemicals can leach out. These chemicals can have a variety of effects on human organs or increase the risk of cancer.

Against this backdrop of effects, what shorthands do we have? The most visible is the recycling code on plastic products, such as "PET 1." This was launched in 1988 by the Society of the Plastics Industry, with the aim of categorizing plastics into different generic types [2.24]. For example, PET 1 is the category that includes soft drink bottles, while PS 6 is for disposable cups. The classification is primarily driven to enable recycling, rather than to label according to the embedded emissions in their production, or the propensity to leach chemicals during their life. Three of the categories—PVC 3, PP 6, and PS 7—are designated as ones to avoid recycling on some broad reasoning, but that is as close a correlation to the full environmental impact as there is. In practice, these codes fulfill a useful role for the waste and recycling industry, but have little benefit in informing consumer decisions. Few people understand the codes, in addition to the codes not correlating well with environmental impact. This leads, in reality, to consumers having some general sense that plastics are bad. This may lead to people simply using as little plastic as possible—reducing *the amount of stuff* used. This may prove to be a surprisingly effective metric.

Aside from our food and houses—and vehicles that we will come to discuss in more detail later—the other important driver of modern quality of life is electricity. It is the force, the carrier of energy, that is produced far away and enters our homes invisibly.

How do we think about that from an environmental point of view? Each unit of electricity as it enters our house is identical. It is a commodity, with each unit being perfectly fungible. Whether that next unit is produced from a low-carbon wind farm or from burning a lump of coal, it is exactly the same electricity in terms of energy and capability. As consumers, therefore, we have no immediate way of assessing how environmentally friendly switching on lights is.

Practically, the only tool at our disposal is to sign up to a suitable "green" tariff from our electricity provider [2.25]. From reputable suppliers, that denotes that some high proportion of the electricity you are buying is from wind or solar if the tariff is "renewable" but can include nuclear too if it is "low carbon." It should be recognized, however, that just because you are on one of these green tariffs, it does not mean that wind-generated electricity is piped directly to your house, or that it is hypothecated in some way for you. As all electricity supplied to your house is from the grid, this is a mix of electricity generated from all sources, including coal and gas. You are trusting the electricity company to buy the electricity from power stations that reflect your choices or, rather than buying electricity from power stations that reflect the overall choices of all their customers. While this could theoretically be verified, and electricity companies publish some data, in practice it has to be taken on trust.

Furthermore, this analysis considers only averages—the environmental footprint of your electricity consumption over a period. It gives you much less information about the source of the marginal unit of electricity when you turn on that light right now [2.26]. Ideally, as an environmentally conscious consumer, you would want to weigh up the utility and quality of life from flicking that switch against the marginal environmental damage. This information would be extremely hard to ascertain in principle and impossible to deliver to consumers in practice. An extreme case of invisible electricity use is operating computers and surfing the Internet. The marginal snap, tweet, or post uses electricity. The Internet in total used approximately 1.5% of total global electricity in 2023 [2.27]. It is not widely realized that there is such a large environmental overhead from the Internet, let alone the environmental impact of each additional minute spent online. There is no shorthand for this currently, and so even the environmentally conscious surfer is unlikely to think about the externality not internalized, and hence the "overconsumption."

The supply-side approach is much easier. We know what fuel is burnt in what quantities in what types of power stations. We can count the coal or gas going in and very accurately estimate the carbon dioxide released from there, using what are called "emissions factors"—standard values, derived from research, that tell you the emissions from each unit of the fuel [2.28]. It is also practical to measure the air pollutant emissions at the top of the chimney, which must be below regulated limits in many countries, made possible by installing pollution "scrubbers," very similar to aftertreatment for cars but called scrubbers in this context [2.29]. Unlike the case of agriculture and houses, where the amount consumed of each may then be the most accurate metric, this does not apply to electricity production. The environmental footprint of a wind turbine or solar panel is mainly in its construction, as it is to a large extent with nuclear. For fossil fuel power stations, the same amount—whether by volume or mass—of the fuel is secondary to the *type* of fuel. Natural gas releases 55% less carbon dioxide per tonne and 48% less per unit energy than typical coal [2.30]. Wind, solar, and nuclear emit nothing and require no fossil fuel to produce the marginal unit of electricity. Therefore, the dominant factor in assessing the cleanliness of an electricity grid is better estimated by knowing the mix of power sources by *type*, rather than by any specific inputs into the system.

Summary

What we conclude from this is that the reality of a world of different types of products and technologies, and the varying ability to collect and distribute relevant environmental data, means that each sector will have different rules of thumb as to environmental impact. Often, these will operate better on the production side rather than on the side of the consumer. In many cases, consumers have less ready access to relevant information and less time to process it, and many have less inclination to. They also must rely on producers and authorities to produce the right rules of thumb. If these are inaccurate, incomplete, or—sometimes—deliberately obfuscating, the environmental objective will be undermined. If too complicated, consumers will ignore or be confused by them. If inaccurate, consumers may confidently make decisions that lead to unintended outcomes. Therefore, as we consider how this applies to vehicles, the two principles should be whether the rule of thumb is *suitable* and *accurate*. "Suitable" in that the rule of thumb chosen correlates as well as possible to the environmental goal in principle. "Accurate" in that the rule of thumb is accurate in practice, and not undermined by missing, inaccurate, or even fraudulent source data.

We started this chapter by highlighting the downsides of simplistic banning of products that are seen as bad for the environment. As always, there are exceptions—limiting cases that are of such clarity or urgency that only a ban makes sense. An example from 1987 is the Montreal Protocol, which banned the use of chlorofluorocarbons in aerosols and refrigerants [2.31]. The science that linked the release of these highly volatile chemicals into the air to the destruction of the ozone layer was as close to conclusive as is reasonably possible. The damaging effects of holes in this layer were well understood, especially the skin cancer risk from ultraviolet sun rays. With both the cause and effect highly clear, with the fast progress of the damage, it led rightly to a rapid ban. A similar case was the banning of lead—strictly, tetraethyllead—in gasoline. It was added to improve the performance of these engines, but by the 1970s, it was well understood to be toxic to humans. Restrictions followed, with a complete ban in the US in 1986 and in the EU in 2000. The last country, Algeria, banned it in 2021 [2.32]. The science was relatively simple and clear, and the intervention was effective.

In 2024, we are racing toward a ban on ICEVs due primarily to their effect on global warming. By the end of this book, you will be able to form your own view on whether the widely believed climate change problem is best addressed with such a ban. Is the shorthand of "banning the ICE" the most effective way to achieve the goal? Are there better alternatives?

3

Doing the Right Thing

IN THIS CHAPTER

- Bans of pollution sources are tempting but hide significant costs and distortions.

- Environmental taxes can influence behavior more than relying on good intentions, and they raise revenue.

- Radical simplification of how we understand the environmental impact of vehicles will allow real environmental improvements from cars and how we use them.

Being environmentally conscious is a form of virtue. It is hard to stick to as it requires either effort or abstinence. Nevertheless, with both virtue and environmental consciousness, most people want to do the right thing, even if they do not always succeed [3.1, 3.2]. As we have discussed in the opening chapters, optimal environmental behavior does not happen automatically because of market failure—we do not bear the full price of our actions and so we overconsume. At a high level, therefore, the trick is to present people with the right incentives so doing the right thing comes naturally. Even when this is successful, though, a price is still paid, whether it is in the form of a "sin tax" or foregone consumption.

The idea of doing the right thing also exists at the government level. Many phrases have been coined to this effect: "green growth," "green jobs," "energy transition," and so on [3.3, 3.4]. All these words are carefully chosen to sound as positive as possible and to imply that there is no reason not to embrace them enthusiastically. They all mean, in reality, in one way or another, that some sacrifice must be made to achieve the outcome. Green growth entails non-green recession, and as most of the economy is non-green—however precisely defined—that will hurt most established economies for a period [3.5]. Green jobs mean non-green sackings. Energy transition means embracing wind and solar at the price of shutting coal and gas power plants and putting coal and gas workers out of their jobs. Each puts a significant price on the green outcome. That is not to say—absolutely not—that the green investment should not happen, but it is not a cost-free exercise. In other words, governments heralding these policies clearly want to do the right thing, but to achieve them will require effort and abstinence as well as support for those affected to manage the transition [3.6].

Economies are never at equilibrium. Some markets within may achieve this state for periods, but for most of the economy, most of the time, markets are in disequilibrium and transition. Resources are being reallocated from where there is too little demand, to where there is growing demand. Jobs are being lost, while new jobs are created. Factories close and new service companies are born. This is the natural state of affairs in a relatively free market economy. Apple Computer grew rapidly off the back of the iPhone launch in 2007, but Nokia and other incumbents shrank as a result [3.7]. Consumers freely switched and their satisfaction—at least in the immediate term—was enhanced. There were winners and losers, but it was a process of creative destruction, as coined by Joseph Schumpeter, that happened automatically via the mechanism of prices and free choice [3.8]. Not only did this process happen automatically, it happened quickly.

The green transition—or whatever term is right to use for a shift to a more environmentally conscious position—is different from a normal economic transition. The power of belief by an individual in whatever environmental cause is not enough to bring about a change at the whole economy level, even if that individual is consistent and persistent in changing behavior. Warm words on environmental topics are not the same as the sacrifice required permanently to change someone's behavior. The nature of the transition required to address our various environmental challenges is more akin to wartime. In wars, more so since the larger twentieth-century wars, whole economies needed to be reshaped quickly to move people from civilian jobs to the battlefield, and industrial supplies into military materiel [3.9]. This could only be achieved by central government direction, either via immediate commands or through bans and rationing. Even where military victory is achieved, the price is high. The most visible price is death and destruction of physical capital. The less visible price is destruction of human and social welfare, and the opportunities for personal growth lost.

Government Intervention

Making the necessary environmental changes will lead to a transition part-way between an automatic economic adjustment and a wartime command economy. It will require government intervention to make a reality of consumers' and businesses' desire to do the right thing. This can be achieved in several ways. Often, the go-to solution for a perceived environmental bad is to ban. Ban single-use plastic bags [3.10]. Ban microbeads in shampoo [3.11]. Ban the ICE [3.12]. Sounds enticing: simple, clear, and easy to enforce. Perhaps they are simple, clear, and easy to enforce, but probably such bans come at a high price. That high price may be a lot less visible than the manifest benefit. Take the single-use plastic bag phased out of many European supermarkets. The apparent benefit is quite clear: plastic bags are gone from the weekly shop. Therefore, plastic pollution in rivers has been stemmed. There are some relatively visible downsides, such as having to pay for a reusable plastic bag or paper bag alternative. The real cost is more hidden, as the environmental damage caused by the production of many alternative bags—whether in carbon dioxide or some other pollutants—is often higher than for the single-use plastic bag [3.13]. This can be related to the production, but equally, it can be a result of a "forever" bag lasting a lot less than forever, making the emissions per use higher than for the thin, single-use variety it replaced.

It would be easy to dismiss this case as an isolated failing, but that would not be correct. In fact, even today, the ban on single-use plastic bags is generally perceived to be a success. So, the perception is wrong, as well as the reality. But did it have to be wrong? While in principle the governments could have cast the ban in a different way to achieve the desired goal, the fundamental problem is that modern economies are complex, and it is impossible for a central government to check and direct every individual production and consumption decision. This could mean checking the production processes of every durable plastic bag factory, or checking that every household looks after its forever bag and uses it to shop in perpetuity. Not practical. Instead, it relies on simple messages communicated to consumers, or businesses, even if they are only blunt tools in achieving the goal. It tends to be the academics and researchers who, after the event, go and look at whether the policy achieved that goal. The rest of this book will look in detail at emissions from transportation, which will give ample evidence as to why a ban on the ICE is a blunt and inefficient tool for achieving transport decarbonization.

Governments, often short of cash, are tempted to use bans as they come at little direct cost to the exchequer. Where a ban is not seen as appropriate or desirable, but taxpayer funds are still short, governments often rely upon persuasion, in the form of official advertising or communications. In many countries, domestic recycling falls into this category. Households are requested to separate certain recyclable items, such as glass and cardboard from organic matter and nonrecyclable items. In most countries, this is not legally mandatory, with exceptions such as France [3.14]. In this case, the hidden costs of the system are much more limited. There are the visible costs of effort and collection costs, but the main issue is participation—do consumers do it on a regular basis, and do they separate their waste in the correct manner? Often, compliance is mixed and can decline over time if the government does not maintain its persuasive messages. If the compliance for domestic waste is patchy on such a minor issue, simple persuasion is not going to be sufficient to transition an economy. Consumers may want to do the right thing in their minds, but this often does not translate to regular practice.

There will be some limited cases where a government educational intervention can both succeed and be sustained. Even though our usual model of the economy is of relatively free agents making choices between a wide range of competitive options, this has limitations. People are often habitual in their choices. Shopping around and thinking take time and effort. Therefore, people are often making somewhat suboptimal choices. In this case, by providing an education and coordinating role, the government could in some cases reduce both environmental impact and improve welfare. For example, if a new public transport service were launched, a "nudge" to get people to try it might lead to a permanent shift, as often it takes a consumer to try a product or service before they can truly put a value on it. Many people are likely to revert having tried it, but those for whom it is an improvement will stick with the new public service. An interesting example of this is permanent changes in commuting patterns after public transport strikes. A 2017 study showed that 5% of commuters in London permanently changed their commute to a faster one after a tube strike forced them to experiment [3.15]. While these effects sound positive, the issue is that these suboptimalities are likely to be at a scale and magnitude to make insufficient difference at the scale of the whole society. They may still be worth doing because of their relatively low cost, but are not sufficient to bring about big reductions in environmental damage.

Struggling to persuade their consumers and businesses, the next step for governments is to use legal bribery, or subsidies. Subsidies can be of any structure or size, but in an environmental context, they are usually thought of as a way to correct for the market failure of overconsumption set out above. If the price of gasoline at the pump is "too low" as it does not include the environmental damage from carbon dioxide emissions, for example, consumers will buy too much gasoline and make too many journeys. To correct this, the government can add a fuel tax of the same magnitude as the environmental damage. This will reduce consumption to the optimal level. Consumers will suffer reduced welfare—their journeys will be more expensive, and they will probably take fewer of them—but tax receipts will be gathered by the government which can be used to offset that loss of private happiness. As with much economic theory, this sounds great but can be difficult to implement well in practice. Due to the complexities of modern economies, there can be many substitute goods that consumers can easily switch to if a tax is applied to their current preference, and collecting the tax receipts in practice without a large deadweight cost of administration is vital. Compromises must be struck between the effectiveness of the behavioral change, the size of the tax receipts, the simplicity of the message to consumers, the distribution and equity effects, and the deadweight and operational costs. Bearing these in mind, such a tax on vehicle fuel is one of the most effective: as the ready substitutes are limited, the tax is clear to motorists and collection is relatively simple through fueling stations.

Many of these comments have been made from the point of view of changing consumer behavior, but the same considerations apply to businesses. In practice, governments often prefer to influence producer behavior, as asking consumers to change risks electoral displeasure. If a business discontinues a product, or subtly changes the recipe for its chocolate bar, many will not notice or care sufficiently to complain [3.16]. Nevertheless, the same concerns for costs and effectiveness remain. When the UK government proposed banning two-for-one deals in supermarkets to try to address obesity, the policy backfired and was postponed because of fear of the effects on poorer households relying on these offers to feed the whole family. The policy would have effectively made food more expensive, in a regressive way that disproportionately affected vulnerable people. Adding labels to food showing calories, sugar, salt, and other key ingredients has been a moderately effective nudge, but not one that registers significantly at the economy level [3.17]. More effective has been "sugar taxes" that have been introduced in several countries, including the UK [3.18]. This has pushed up the price of many products including confectionery or reduced the size of the chocolate bar for the same price. This is seen to have had a positive effect in reducing obesity, even though the price is borne in terms of reduced consumer economic welfare and potentially producer profit. To reemphasize, even effective schemes to change behavior come at a cost, even if the cost is quite hidden.

All in all, there is much evidence that both consumers and businesses want to do the right thing for the environment, but this is not enough to bring about automatic and sustainable behavioral change on its own, and central interventions can easily be either ineffective or expensive. This can be seen clearly with consumers and their cars. It used to be much easier than it is today. Switching from horses to motor cars in the early twentieth century was such a clear improvement, as no amount of carbon monoxide and carbon dioxide from early gasoline cars could be worse than mounds of manure gathering on the road, with the associated smells and diseases. After World War II, physically small, cheap cars brought motoring to a wide population, which were ostensibly cleaner and more efficient than the big luxury cars of the upper classes. Los Angeles was the first major city to see the air quality impacts of the democratization of motoring [3.19]. As most cars ran on gasoline, engine size was a good way to estimate the environmental impact. Then came diesel as a fuel for passenger cars. Although France and some other European countries had been pioneering diesel cars for decades previously, the big push in Europe came in the 1990s as the first fleet of carbon dioxide targets were introduced [3.20]. Diesel was the answer governments envisaged, as it operates approximately 25% more efficiently than gasoline, so despite its 13% higher carbon content, it releases 12% less carbon dioxide per mile or kilometer driven [3.12]. Twelve percent was a big potential reduction and one that could be achieved relatively easily as diesel fuel was already available through the refueling network. It was easy—according to official data at least—to do the environmentally responsible thing by buying a diesel.

Dealing with Complexity

Then came Dieselgate in 2015. We will discuss this in more detail later, but at this point, it is sufficient to consider its implications.

The reputation of diesel, and the automotive industry as a whole, was profoundly damaged. Most consumers buying new cars switched back to standard gasoline ICEVs, and a few to gasoline hybrids such as the Toyota Prius, originally launched in 1997. Then, Tesla launched its first mainstream car, the Model S, in 2016—a pure BEV. Sandwiched between pure electric and hybrid vehicles were various "plug-in" hybrid vehicles with both an engine and a larger battery [3.12]. The different types of powertrain—the combination of engine and transmission—proliferated. Many involved innovative technology. The world of buying a car became complicated. As an environmentally conscious consumer, which should I choose? Surely, it is the pure battery vehicle? Or maybe the simple hybrid. Hmm…

Complexity for consumers is a recipe for confusion and suboptimal choices. If I do not have the data, or do not understand it, or do not trust it, I am likely to revert to some prior rule of thumb, or urban legend, or prejudice. I might get bored and choose based on color. As we have seen previously, consumers generally do not have the necessary expertise or time to research every option in detail, and nor should they be expected to. Central authorities should oversee the good functioning of markets and ensure that producers—or some other body—provide reliable and digestible information to inform decisions. Today, we are stuck in relative informational poverty and consumers are making semi-informed choices. And producers have to second-guess what consumers want, as well as the future direction of policy.

This problem is what, in essence, this book will aim to solve. How can we make buying a car, and doing the right thing environmentally, easy again? To make this work, consumers will need access to simple, trustworthy information at the point of choosing candidate vehicles and then making the ultimate purchase decision. This proposal is unlikely to be controversial in principle, yet hard to deliver in practice. Prior to 1992, there were no official data on vehicles, so environmentally conscious consumers used available proxies from the mechanical specification of the vehicle, such as engine size or fuel type. With the introduction of the "Euro" emissions standards in that year, European consumers then, in theory, could consider the extra piece of information that any new car on sale should emit no more than the legally defined limit—or "limit value," in the jargon [3.21]. Even with this very first stage of emissions regulations, there were complexities. The most obvious was that multiple pollutants—including carbon monoxide and total hydrocarbons—were each measured against separate limit values and not combined into a single rating. Therefore, consumers who did consider this information had a lot to juggle in their heads. Fortunately, at least, they were shopping in a more trusting and naïve age where official data was believed and trusted. The divergences between official data and reality only became clear later, culminating in Dieselgate [3.22].

The natural response to the problem of multiple pollutants is that they should be combined into a single "green" rating. This has never been done officially but was done privately in 2018 with the launch of "Green NCAP" ratings, funded significantly by the FIA Foundation, the charitable foundation of the governing body of motorsport [3.23]. Cue sighs of relief from consumers? The approach of combining pollutants, which are inherently different in their environmental and health effects, is logically problematic. It is necessary to weight each pollutant into the overall average—but what should the weightings be? The obvious answer is to weigh by the severity, but this itself is complex, multidimensional, and, to a great extent, subjective. For example, someone living on the polluted streets of a poor part of London may be much more concerned about nitrogen oxide emissions from old diesels than carbon monoxide emissions from expensive classic cars. What these singular ratings give is a false sense of knowledge, as the consumer will rarely read and cogitate on the assumptions set out in the extensive methodology document.

Such a voluminous ratings methodology is a testament to the amount of work and resources that has been put into devising the Green NCAP rating system, and the technical credibility of the testing work that powers it. But the devil is in the detail. The most glaring example of how the choice of which pollutants are included, and how they are weighted, affects the overall result is Green NCAP's choice to exclude carbon dioxide emissions in vehicle manufacturing from the rating. There are reasons for this—flowing from the availability of the necessary data from car companies—but the outcome is that all BEVs gain an almost perfect score! Not because they are almost environmentally perfect, but because of the choice of how to construct the ratings system. An artifact, in the jargon. Therefore, Green NCAP works to give either false guidance to the casual consumer or a huge amount of homework to the diligent consumer. Nothing quick, easy, and reliable here.

Tax

All this discussion through the lens of the consumer can also be seen from the governments' fiscal perspective. Vehicle tax regimes around the world are a complex, sprawling mess of rules brought in over time for various reasons from the environmental well-meaning to the knee-jerk reaction to the latest crisis. For example, the taxation system in the UK includes an *ad valorem* tax (a tax based on the value of the transaction) of gasoline and diesel at the pump, on top of which there is also value added tax (VAT). In the early 1990s, a "fuel duty escalator" was introduced with the aim of giving forward visibility of future increases in the level of fuel duty, motivated at least in part so the environmental damage from these vehicles was increasingly borne by motorists creating the pollution. However, in 2003, in response to spiking crude oil prices, the government suspended the escalator [3.24]. It did not abolish it, but suspended it. The suspension has been rolled over, year by year, most years, by successive governments of different shades of political opinion such that the fuel duty has fallen in real terms throughout the twenty-first century—the 2023 value is approximately equivalent to the 1993 value when adjusted for infla- tion and has fallen every year since 2011 (see Figure 3.1)—effectively an increasing subsidy for drivers for over a decade [3.25]. This is a good example of taxation with a good environmental motive that gets bastardized by events and political decisions. What was policy designed to give forward visibility on taxation rates became a glorious example of perennial uncertainty—each year we speculate as to whether the fuel duty escalator will be unfrozen.

FIGURE 3.1   Evolution in real terms of UK fuel tax, 1989–present [3.25].

Duty in real terms

Pence per liter 2022 prices

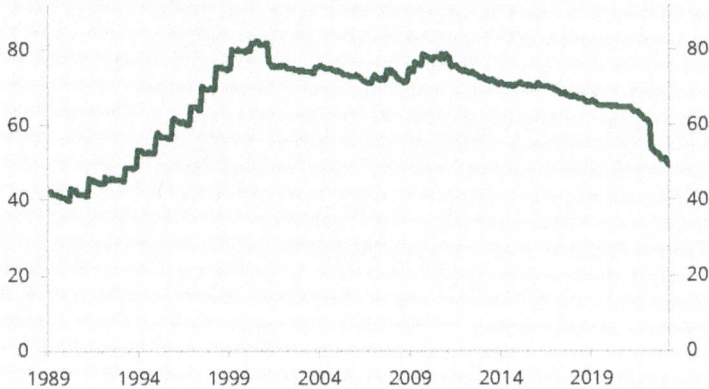

And this is only one example. The fuel duty escalator sits within a tangle of other vehicle and fuel regulations. The exact situation varies from country to country, but most developed countries have a similarly convoluted patchwork of taxes and rules. And yet now, the total tax take is falling, driven primarily by reduced fuel duty. This is not yet primarily because of more BEVs on the road—they only make up 2% of cars on UK and European roads—but from the increasing fuel efficiency of ICEVs and a switch to various forms of hybrid vehicles [3.26]. So, the tax system is a mess, and revenues are declining. A re-evaluation, a reset, is needed. Good taxation works by being simple and effective. The current system is complex and increasingly ineffective. Good taxation must also maximize the intended market distortion and minimize the unintended distortion, all while maximizing the revenue it brings in.

The second hypothesis of this book, therefore, is that the same approach to simplifying consumer choice for the environmentally conscious can also be the way to reform the taxation system in a way that increases revenues and delivers environmental policy goals, while minimizing undesirable distortions to free economic choices.

Paradoxically, the error on both the consumer and government side may have been that, in trying to incentivize the reduction in emissions by tying information and taxes to those emissions, it becomes self-defeating. This is not as peculiar as it sounds and reflects effects that are seen elsewhere, such as Goodhart's law, which says that any economic indicator ceases to measure what it was designed to measure as soon as it is targeted in policy [3.27]. In the context of emissions reduction, we can see many examples. The most lurid one, which will be discussed more later, is Dieselgate, where official figures for nitrogen oxide emissions underestimated real-world emissions by approximately 80%, as a result of vehicle manufacturers in Europe optimizing their certification testing to the letter of the law. Whether this will prove to have been unlawful, beyond a limited number of cases, is still being tested in court. However, during the same era, a distortion of official figures was also being performed by manufacturers for carbon dioxide emissions. The UK vehicle excise duty system—an annual vehicle tax—was based on carbon dioxide emission bands of typically 10 g/km. For example, a vehicle emitting between 200 and 209 g/km would attract a lower tax than one emitting 210 to 219 g/km. Observing the declared emissions from manufacturers found a significant clustering at the high end of bands, for example at 209 and 219 g/km [3.28]. In other words, manufacturers likely performed a combination of designing their vehicles and optimizing to the official test in such a way as to just creep into a lower band. In other words, the effect of the tax was not to target lower emissions per se but to optimize according to the system. A tax regime ostensibly—and legitimately—designed to internalize the externality of carbon emissions was subverted.

The underlying issue here is that too many of the players in the system have the incentive to subvert it. Diligent consumers want the truth, but vehicle manufacturers want to flatter their vehicles and government regulatory authorities want to create a system that either flatters them, produces a popular answer, or panders to predominant external interests. In Germany and France, this has typically been to create systems that are flattering to or can be used to good advantage by their domestic automotive industries.

On the surface, the US federal regulatory system has been better at creating official emissions figures that reflect reality, but the distortions emerge elsewhere. In the US case, it is in the resistance to incorporating particle number emissions—which matter as these represent potentially more dangerous ultrafine particles—into the official reckoning [3.29, 3.30]. While the evidence as to their danger is not yet complete, the EU regulated the precautionary principle anyway.

To tackle this underlying tendency of systems to subvert themselves, our hypothesis pivots around whether we get back to a simpler world where simple-to-verify mechanical information can be used as a proxy for a car's environmental impact. Perhaps, there is an analogy with domestic waste. It is relatively straightforward for households to put cardboard and plastics in one bin and food waste in another. The motivation is for the purposes of both recycling and emissions reduction, for example by reducing methane emissions as food decays in landfills. Households are not being asked to segregate waste on some emissions or environmental numbers, but on simple, observable facts about the item being discarded. Could that work for automotive? If so, what would the separation criterion or criteria be?

If we do not tackle this urgently, there is a significant chance that the move to electrified vehicles will set up a crisis much larger than Dieselgate sometime next decade. Environmental regulation of vehicles targets primarily in-use emissions, either directly via taxes on emissions or semidirectly on fuel purchased and burnt. BEVs, quite obviously, have no tailpipe emissions. As they can be charged up from domestic electricity supplies, it is also a significant challenge to tax the electricity usage for cars higher than for other domestic uses. The more important emissions from these electric cars come from the manufacturing process—mining and refining of materials that go into the batteries and motors. The added weight and torque—the force from the motor that generates the acceleration—of battery vehicles compared to their internal combustion equivalents lead to much higher tire wear emissions. These emissions are potentially more toxic for humans and animals because of the complex chemical cocktail that tires are made of.

As it stands, neither the "upstream" emissions from vehicle construction nor these tire emissions feature at all in environmental taxation. Resting taxation solely on in-use emissions and fuel use will create a distortion that will lead to lower social welfare and higher environmental impact than is optimal. The potential scandal coming may arise from omission—not changing the system to reflect current realities—and from influencing the structure of the emerging "lifecycle" models designed to incorporate all these various elements.

Summary

Without a radical simplification in how we think about the environmental impact of vehicles, we will continue to descend into a fog of confusion. We are assailed by many environmental initiatives, which are often positioned as panaceas. In a sense, these are being put forward as the routes to simplification, but invariably the marketing does not reflect the scientific reality. Take the Ultra Low Emissions Zone (ULEZ) launched in London in 2018 and expanded in 2023. If you want to drive a pre-Euro 4 gasoline vehicle or a pre-Euro 6 diesel vehicle within Greater London, you must pay a daily fee, now £12.50 (approximately $15) [3.31]. The broad policy goal is to improve the quality of London's air. The specific goal is to reduce nitrogen oxides emissions, of Dieselgate fame. The Euro 6 cutoff for diesel vehicles aligns fairly well with this motive, but much less well with the gasoline restriction. Gasoline vehicles, because of the nature of spark ignition combustion, generally do not have a problem with nitrogen oxides, but do with carbon monoxide and unburnt hydrocarbon emissions as their catalytic converters age and deteriorate. In addition, gasoline engines with direct fuel injection into the cylinder (GDI) tend to have higher particle number emissions, despite being a more recent technology (diesel vehicles do not have this problem since Euro 5 because of the fitting of very effective exhaust filters). Nevertheless, the Euro 4 is not a bad cutoff as a proxy for this vehicle aging.

All these vehicles, which contain ICEs, will still emit significant carbon dioxide in use, whether before or after the cutoffs. So, when certain representatives of the London authorities attempt to project the ULEZ as contributing to climate change mitigation, they deviate from the truth and the science. In attempting to project a more universal value in the ULEZ, no doubt to claim political credit, a false promise is made to consumers. Driving a Euro 5 rather than a Euro 3 gasoline vehicle will do nothing to reduce climate change. In other words, the ULEZ access criteria cannot fulfill the role of a simple yet universal metric for the environmental profile of vehicles. These promised panaceas are often little more than adding a badly designed tax on motorists, creating distortions and reduced social welfare. A better approach would be to develop a simple, universal metric to price the environmental cost of motoring and then allow free choice to work its magic.

The rest of this book will pursue this quest.

4

The Taxation Mess

IN THIS CHAPTER

- Vehicle taxation globally sees a wide variety of systems, but few countries tax cars based on their weight.

- Most jurisdictions do not tax, or tax very lightly, BEVs, and this is unsustainable.

- Vehicle tax revenues are declining in many places due to better fuel economy and the transition to BEVs, and these tax revenues need to be replaced.

The history of taxation consists mostly of kings and queens raising revenue to fight incessant wars, often facing internal rebellions when they overdid it. Such taxes were often arbitrary. However, governments did not solely raise taxes to fund themselves, or at least not straightforwardly. With the rise of democracy, broad-based affluence, and new technologies that both stimulated the economy and allowed the newly affluent to express their wealth—the automobile was the quintessential twentieth-century example—taxation became, in the main, less arbitrary as it came to have democratic legitimacy. It meant that, in the awkward jargon of taxation, it came to be "parametric," or governed by legislation defining, in advance, the tax base. That means that we can all calculate in advance what our tax bill will be before it arrives.

In theory, this is also true for car taxation, but in practice, for ordinary citizens, it is not. There are so many competing and overlapping taxes that it has become a contradictory mess, poorly aligned to much-trumpeted environmental goals and offering consumers a poor sense of where they stand. The UK is a leading example of this automotive taxation mess, which forms part of one of the longest tax codes in the world, running to thousands of pages. Let us take a further look at the UK situation.

UK

The first tax is vehicle excise duty (VED), which is applied to vehicles based on their price when new; this type of charge is known as an *ad valorem* tax [4.1]. More recently, a supplementary, progressive, element was added for new vehicles sold for more than £40,000, which, while technically still an *ad valorem* tax, was really a surrogate wealth tax. Simultaneously, the tax is also dependent on some combination of carbon dioxide output, fuel, and vehicle powertrain (internal combustion, hybrid, or electric). Then, there is a secondary raft of "benefit in kind" taxes paid by consumers on cars registered by businesses [4.2]. Further, there are complex excise duties applied to different fuels, different treatment of electricity and gasoline, and yet other tax incentives for alternative fossil fuels like liquified petroleum gas (LPG) [4.3]. VAT hovers over all the above, but not in a coordinated fashion—it is applied at a much reduced 5% rate for domestic energy, benefitting already affluent homeowners with home charging for BEVs, while excluding people who live in blocks of flats [4.4]. The UK charges tolls on a few elements of infrastructure but not on most. Some bridges attract a fee, and there is one toll motorway called the M6 Toll [4.5]. As you can see, no normal consumer would realistically be able to fathom in advance the total vehicle taxation they would pay when they choose a car to buy.

The taxation of cars across the rest of the world is a riot of different systems, and no two systems are the same [4.2, 4.6, 4.7]. Typically, countries charge a one-time registration or purchase fee, plus import duties, VATs, excise taxes, additional registration fees, and environmental taxes. In most countries, there is a recurring annual tax of some kind. Driving, the world over, is a licensed and taxed privilege.

In the broadest sense, there is an observable pattern where oil-producing states impose smaller taxes and tend not to encourage BEVs and hybrids [4.2]. Meanwhile, countries that are heavily reliant on vehicle imports tend to levy larger import duties, sometimes to compel local assembly, if not comprehensive manufacturing, of vehicles. There are standout examples like Singapore where the initial registration tax can almost triple the price of the vehicle being acquired. Other countries like Israel and Japan have complex, high-tax systems for vehicles, while Iran and Kuwait, for example, do not [4.2, 4.8].

Tax revenue from private mobility is set to remain a key source of general taxation [4.9]. Governments tend to apply taxes where the elasticity of demand or supply is low, so behavioral change is limited. This minimizes the economic distortion. Most households consider that they need a car. This relative inelasticity is in apparent tension with the other motivation for vehicle taxation, which is to apply a price to the negative environmental externalities from driving. In this case, there is a deliberate desire to influence behavior, and that behavior can be sensitive to the level of taxation. In reality, though, these two motivations are not necessarily in tension as both tempt governments to set vehicle taxes high: people will buy the cars anyway, but at the margin, they may drive fewer miles.

The UK offers ample evidence of how vehicle taxation is a lucrative source of revenue, and governments act to optimize that. The seventeenth-century French finance controller Jean-Baptiste Colbert stated that: "The art of taxation consists in so plucking the goose as to obtain the largest possible amount of feathers with the smallest possible amount of hissing." Today, vehicle taxation goes toward paying for many areas of government spending, including healthcare. There have been times in history when vehicle taxes have been hypothecated toward (reserved for) road spending, and there has even been recent discussion about returning to some version of that in the UK [4.10], but in reality it is a too tempting and fertile source for covering wider government spending. Transport has become a big revenue earner for governments the world over.

Although BEVs initially attracted preferential treatment and were zero-rated for VED in the UK, the government, seeing that it was losing revenue and potentially staring at a £35 billion hole in revenue by 2050 if net-zero goals were met, moved the goalposts. In 2022, it was announced that BEVs would no longer be exempt from VED from 2025 [4.1].

Sticking with the UK example, that revision from 2025 does not address the broader problem, of how to plug the huge hole in—particularly—fuel duty that will rapidly emerge if drivers switch to BEVs. The tax revenue from fuel duty and VED tax together account for 4% of the entire tax revenue in the UK [4.11, 4.12]. Today, although many people do not realize it, most taxation on vehicles is directly or indirectly a pollution tax. Fuel taxation (so-called gas tax) is the most obvious example. Burning 1 L of gasoline will create approximately 2.35 kg of carbon dioxide, and 1 L of diesel will create approximately 2.7 kg of carbon dioxide (the equivalent numbers for 1 US gallon are 8.9 kg of carbon dioxide for gasoline and 10.2 kg of carbon dioxide for diesel). It does not matter what type of vehicle you burn the fuel in (or whether you burn it in a vehicle at all), the carbon dioxide emitted from burning the fuel is driven overwhelmingly by the carbon content in that fuel.

In the UK, the fuel duty, which is set at 52.95 pence per liter (p/L), is equivalent to a tax of 22.53 p/kg for gasoline and 19.6 p/kg for diesel ($0.274 and $0.239/kg in US dollars, respectively—using the exchange rate in October 2023) [4.12]. In the US, the situation varies by state and taxation is a combination of local, state, and federal taxes. The highest-taxed state is Pennsylvania, which taxes at 76 cents per gallon (c/gal) for gasoline and 98.5 c/gal for diesel, and Alaska has the lowest rate, which taxes at 27.35 c/gal for gasoline and 33.35 c/gal for diesel. This corresponds to $0.0854/kg of carbon dioxide for gasoline and $0.0966/kg of carbon dioxide for diesel in Pennsylvania, and $0.0307/kg of carbon dioxide for gasoline and $0.0327/kg of carbon dioxide for diesel in Alaska. In both cases, we have excluded sales taxes and VAT from these figures, which are levied on the fuel duty or tax as well.

These contrast with non-transportation carbon dioxide taxes, such as the EU's emissions trading scheme, where carbon is traded at approximately €85/tonne of carbon dioxide at the time of writing, which would correspond to $0.0896/kg of carbon dioxide. This is roughly in line with the highest US taxes and much lower than the UK taxes.

While the idea of taxing the negative environmental externalities is logically strong, governments can make a complex mess of it too, as has happened in the UK. The dependency of VED on carbon dioxide emissions was originally introduced to incentivize the uptake of more efficient diesel vehicles in 2001 [4.1]. This was successful, but regulations and taxes across Europe—not just in the UK—neglected the associated air pollutants of nitrogen oxides, which led to the Dieselgate scandal and declines in air quality [4.13]. In reaction, a diesel premium was added to the VED regime in 2018 for vehicles not meeting the very latest nitrogen oxide emissions standards called "RDE2" [4.14], to counterbalance the original incentives in favor of diesel.

Global Context

Like general taxation, vehicle taxation is colorfully different around the world. No two systems are alike [4.2, 4.6, 4.7]. In most theories of globalization, there is an invisible march toward harmonization, and one might expect that to prevail above all in taxation, where the balancing act is between three fundamental objectives: tax revenue, equity, and efficiency. In 1960 the only country in the world to have VAT was France. By 1980, the number was 27. By 2019 the figure was 166 [4.15], appearing to back Adam Smith's point that "There is no art which one government sooner learns from another than that of draining money from the pockets of the people" [4.16].

Yet, that sort of convergence has not happened in the area of transportation, mostly, we suspect, because of particular historical and social experiences and different geographies, plus the impact in recent decades of environmental objectives over and above traditional considerations. The latter has added to astonishing complexity and, in some cases, resulted in unintended and occasionally downright bizarre outcomes, as with the UK and Europe and its relationship with diesel vehicles.

To seize one example of a historical incident that still exerts its pull nearly a quarter of a century later, Britain saw a transport "rebellion" in 2000 around fuel tax, led by haulers and farmers blocking fuel refineries [4.17]. This is one reason why a fuel duty "escalator" has in practice not been activated for year after year by British politicians, as discussed in the previous chapter. This matter is an enduring source of political fear, exemplified by the yellow vests (*gilets jaunes*) movement in France that began late in 2017—an even greater social protest arguing for lower fuel duties in the face of rising crude oil costs, among other, less focused grievances [4.18].

We can spot common threads across national boundaries in these two examples, but the reality is that Britain and France tax their transport sectors very differently, as France's recent move toward weight-based car taxation shows—an example we will turn to. Other countries apply very different principles for their approach to car taxation, and in a handful, there is a weight component.

Could a weight tax be an efficient way of reinforcing the tax base, internalizing the environmental externalities, and minimizing the goose-hissing? Some countries have been dabbling with it [4.6]. Transport & Environment, a lobby group, has suggested that a reduction in the size of cars could reduce the predicted requirement for battery resources, such as cobalt, lithium, nickel, and manganese, by up to a quarter [4.19]. They are advocating for the UK government to impose a taxation system based on weight for the heaviest cars and the largest models, as well as introducing tax incentives for small BEVs.

Such policies are not new when viewed globally. Size-based taxation aimed at keeping congested city streets unclogged is a celebrated feature in Japan, which has fostered an entire "kei" class of vehicles that are made solely for the domestic Japanese market and must not exceed 3.4 m in length or 660 cc in engine capacity [4.20]. The average weight of a non-electric kei car made over the past two decades has ranged from 700 to 850 kg, versus a global average in the 1970s of 1 tonne rising to a global average in 2023 of 1.4 tonnes [4.21]. The nearest European example is the original Smart Car from 1998, which was 2.5 m in length and had a weight of just more than 800 kg, and its successor, the Smart Fortwo, which weighed 720 kg [4.22]. Japan also has a weight tax, which favors smaller, lighter vehicles [4.23]. Unusually, it features as part of the recurring, annual car tax rather than only featuring in the initial registration fee of the vehicle.

Weight used to play the leading role in car taxation in Hungary, but was phased out in 2009 in favor of a performance-based tax [4.24]. Weight plays a role in car taxation at vehicle acquisition in several jurisdictions now, including Australia and the Philippines [4.25, 4.26]. In India, a blend of factors determines the VED, including engine size, seating capacity, and price [4.26].

Latvia has a blended approach to car registration tax that includes weight [4.2, 4.23]. For passenger cars registered before January 1, 2005, road traffic tax is calculated on maximum gross weight in kilograms. The lowest band is a gross weight of below 1500 kg, costing €38. The highest is 3500 kg and above, costing €274. For vehicles registered between 2005 and 2008, the metric was expanded from weight to include engine capacity and power. And then, in a commonly observed adjustment, from January 1, 2009, traffic tax was calculated by carbon dioxide emissions per kilometer, replacing the previous systems. For light and heavy commercial vehicles, weight continues to play a leading role in taxation in Latvia.

The recurring ownership tax in the Netherlands is assessed on a blended measure of gross vehicle weight, fuel type, region (province) of location, and carbon dioxide emissions. Road tax has also been assessed since 2016 using the same elements, with BEVs being granted an exemption from 2021 to 2024 [4.2, 4.23].

In fact, if you look around the world, most governments have opted for a blended approach that looks at the fuel type of the vehicle, its power output, and its sticker price as the basis for registration and recurring duties. This blended approach is unlikely to change. The reticence toward moving to a weight-led approach arises from a potentially justifiable fear—that it would slow the BEV revolution because these vehicles are generally heavier than their ICEV counterparts [4.27]. There are some signs of a new, dual approach, which penalizes the purchase of heavy cars but takes a more lenient approach to BEVs without exempting them completely from a weight-based approach. We will now look at four countries in further detail to explore how they are doing this.

Norway

In Norway, BEV subsidies have been so generous that in 2022 they reached 2% of the national budget, amounting to $4 billion. This has now caused some degree of anguish, with the authors of an Organisation for Economic Cooperation and Development (OECD) report noting how this has been distributionally divisive by subsidizing the already affluent and distorting the investment case for alternative rail and bus solutions [4.28].

Of course, we should not neglect the success of Norway in rapidly achieving a modal shift toward BEVs. In 2011, only 1% of new vehicle purchases in Norway were BEVs [4.29]. By 2022, the number had risen to 79% [4.30]. There is no other jurisdiction in the world that comes close to this level of BEV penetration. BEVs comprised just 3% of car sales in Japan in 2022, for example [4.31].

The average sales share of BEVs in the EU is 15% [4.32], and in the US it remains 8%, in 2023 [4.33].

However, the mass adoption of BEVs in Norway has also created unforeseen problems. We have already mentioned the regressive, distributional problem, which has tended to reward affluent city dwellers at the expense of the rural poor. But the other unforeseen problem in Norway has been the surge in outsized vehicles that are hard to describe intuitively as green, if only for their enormous bulk and weight (the average vehicle weight in Norway has climbed from 1270 kg in 2000 to 1970 kg in 2023), quite apart from the implications for road wear, road safety, and unsuitability for narrow city streets [4.34].

This led the authorities to introduce a weight element to vehicle taxation in January 2023, as shown in the chart below [4.35]. It is punitive for ICE-powered SUVs, but still there, in the mix, for BEVs at a much lower level, aimed at reducing sales of the largest battery electric SUVs, such as the Hongqi e-HS9 SUV, weighing in at approximately 2660 kg or 5900 lb [4.36].

FIGURE 4.1 Norwegian weight tax proposal [4.35].

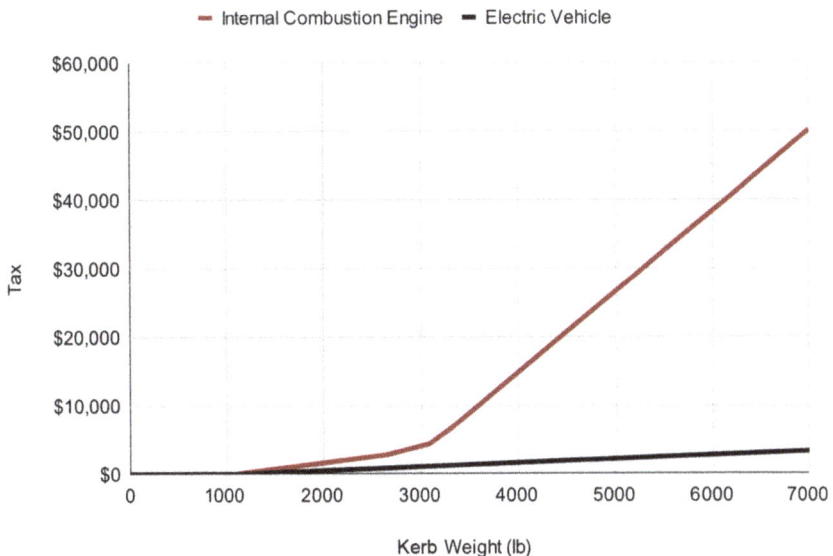

Purchase Tax on Passenger Vehicle Weight (Norway 2023)

Courtesy of Matt Bruenig.

Call it bonus/malus or "feebate," or let us call it carrot and stick, this is about how to nudge or cajole consumers and the industry toward making and buying some vehicles more than others, toward a broader goal that is typically now connected to climate objectives. A more detailed example of this, again implementing a weight tax, is in France, which we will turn to now.

France

France is a classic tale of tax complication, but also a working example of how implementing a weight tax might work despite its detractors trying to stop it. The narrative begins in October 2020, when the government introduced a weight component proposal to vehicle taxation to try and arrest the inexorable march toward ever heavier and perceived less environmentally friendly cars.

This government proposal was to levy a fee of €10/kg above 1.8 tonnes to discourage the purchase of heavy vehicles [4.37]. Of course, it was not quite as straightforward as that sounds—it rarely is. Firstly, the French legislators set out complex exemptions for large families and for electric, hydrogen, and hybrid cars, although they are often the heaviest vehicles. Then, a constitutional challenge was made by some lawmakers. Responding, France's Supreme Administrative Court (*Conseil d'Etat*) dismissed the challenge on December 28, 2020 [4.38].

The Conseil's defense of the weight tax is worth noting here. It argued that

> ...the difference in tax treatment provided for in the proposal is justified by the legislature's intent to discourage individuals from purchasing larger vehicles that cause environmental harm by consuming more energy and material and by occupying space in traffic [4.38].

The lawyers opposing the new weight tax argued that the principle of equality was damaged by the fact that the 1.8-tonnes threshold exempted most vehicles in a country where average vehicle weight is still 1.4 tonnes, which is lower than the OECD average, despite being 40% higher than the global average in 1970 of 1 tonne [4.21]. They also argued that "the exemptions for electric, hydrogen, and hybrid vehicles are not justified by the objective to tax vehicles by weight" [4.38]. In other words, how can a weight tax not apply to the heaviest vehicles? We agree that this is an important consideration.

The objectors even argued that the bill was "confiscatory" in nature precisely because it would discriminate mostly against foreign-manufactured vehicles, which may have an echo of German lobbying, or at least a suspicion of French protectionism—with many larger, premium SUVs being made by German, Korean, Japanese, or Chinese manufacturers.

In the end, all the arguments were rejected and a year later, on January 1, 2022, the *Taxe sur la masse en ordre de marche*, better known simply as the *malus de poids*, came into effect. The tax means that for every 100 kg above the 1.8-tonne threshold, France adds €1000 in registration taxes, a one-off payment only due for vehicles registered from that date. That translates into over €4000 extra payable on a 2024 model Audi SQ8, for example, which weighs 2240 kg, although it would not cause any additional bill for the buyer of a 2024 Kia EV9, which weighs 2625 kg.

On June 26, 2023, French Transport Minister Clement Beaune announced that the policy would be continued into 2024 and even extended to cars lighter than 1.8 tonnes, in effect lowering the threshold—yet still completely exempting BEVs [4.39].

Complementing this, although also to some extent dissonant, was the vote in Paris in February 2024 to increase parking charges on heavy vehicles—billed as targeting SUVs. Although the margin of victory and turnout were low, this will progress to increase the hourly parking charge in central Paris to £18 for heavy vehicles. The dissonance with France's other weight-based taxes is that the threshold above which the higher charges kick in is 1.6 tonnes, rather than the 1.8 tonnes above. Furthermore, there is a leniency hardwired into this for BEVs, as the elevated charges only apply to such vehicles over 2 tonnes in weight. Nevertheless, this scheme does add to France's overall move toward charging heavy vehicles more [4.40].

US

The US is inherently complex for the purposes of car taxation, partly because of overlapping tax jurisdictions: federal, state, and, sometimes, city and county. If you buy a car in Los Angeles, the total tax and "tag" (registration) fees comprise 14 different hypothecated subdivisions—right down to a $1 freeway emergencies fee and a $1 fingerprint ID fee [4.41]. But at least the subdivisions declare their purpose. Over in New York state, we find evidence of perverse cross-incentives. There is already a weight-based vehicle use tax there, which particularly penalizes vehicles heavier than 3500 lb (1588 kg) [4.42]. Although this varies slightly across the 28 counties in the state and is typically only a few dollars, the principle is there: you pay more tax if the vehicle is heavy. Meanwhile, however, the Federal tax agency, the Internal Revenue Service, has pulled in the opposite direction, making it cheaper in some instances to own a much heavier vehicle. The Tax Cuts and Jobs Act (TCJA) temporarily allowed 100% expensing for business property acquired and placed in service after September 27, 2017. The arrangement began to taper in 2023 and will expire on January 1, 2027. While it lasts, the first-year tax deduction incentivizes buying a heavy SUV, pickup, or van instead of a luxury car of the same sticker price provided that the vehicle weighs more than 6000 lb (2720 kg) [4.43]. We spotted a Range Rover dealership in New York that at the time of the original arrangement boasted:

Since the Range Rover, Range Rover Sport, Land Rover Defender and Land Rover Discovery have Gross Vehicle Weight Ratings (GVWR) greater than 6000 pounds, they can be fully depreciated in the first year of ownership when used for business 100% of the time, giving you the freedom to spend on the things that matter most [4.44].

The tax savings amount to thousands of dollars while the preexisting county weight tax is normally a matter of just $5 or $10 a year. Here, we therefore see two different taxing authorities aiming at different outcomes and arriving at a contradiction.

Japan

Japan remains a compelling example, even though in recent years the government has reduced the benefits once associated with the kei class of small, light vehicles, described previously. The kei class, which began in 1949 and means "keijidosha" or "light vehicle," expanded to BEVs in 2009 when Mitsubishi introduced the i-MIEV, subsequently sold in other jurisdictions including Europe.

Japan and its carmakers have been reluctant to embrace electrification for these very reasons, and Toyota famously pursued hybrid technology in preference to pure electrification and continues to stick largely to this position [4.45]. It is faced with this conundrum that Japan has in recent years watered down some of the tax benefits that kei cars used to attract—in their ultra-light combustion engine form, it is almost as if they have been too successful, while being difficult to convert to convincing electric versions owing to battery weight.

But Japan has not gone all-in on BEVs like the French. Japan has an environmental performance tax that is a blended measure with weight, age, and fuel efficiency components. It benefits new drivetrains like the BEV, but it also rewards alternative fuels and recognizes super-clean combustion drivetrains [4.23]. The system offers rebates for BEVs, fuel cell electric vehicles (FCEVs), plug-in hybrids, but also liquified natural gas and clean diesel, in the form of reduction in or exemptions from tax where some combination of fuel efficiency or emissions are met. The special treatment of

BEVs in Japan is still there. Vehicles that beat the 2020 Fuel Efficiency Standard by 40% or more "are exempt from a weight tax" [4.26].

Scandinavia and Others

A common pattern we find is a weight-based tax two or three decades ago that gave way to a fuel efficiency basis, which in turn was superseded by carbon dioxide. Denmark is a case in point. Heavier vehicles are subject to higher registration taxes, but only if registered before July 1, 1997. Fuel efficiency then took over from weight, and later on, carbon dioxide replaced fuel efficiency [4.23]. This last modification is only significant when alternative fuels are on offer, as fuel efficiency and carbon dioxide emissions are almost perfectly inversely correlated for the same fuel used. Comparable changes were made in the Netherlands and in Sweden [4.23, 4.46].

Finland is of interest because it has maintained two kinds of weight-based tax: a "basic tax" and a non-gasoline "power tax," which penalizes diesel and incentivizes BEVs, both with a weight component [4.23]. For the basic tax, where emissions data are not available, total mass is considered to reflect the fuel consumption of the vehicle—and total mass data is available on all vehicles. If the total mass of a vehicle divided by 100 kg and rounded upwards is m, and the carbon dioxide emissions value is h, the basic tax can be calculated using the formula: $h = 10 \times m + 7$. For the power tax, all pure gasoline cars are exempt, and then there is a daily fixed charge for every 100 kg of vehicle mass. Rates vary, from 0.5 euro cents (0.53 dollar cents) per day at the low end, for gasoline hybrid cars, to 5.5 euro cents (5.9 dollar cents) per day at the top end for diesel cars.

Summary

In general, therefore, where there are elements of taxation based on weight—and these are gradually becoming more common around the world—there are almost always blanket exemptions for BEVs. This is unlikely to be sustainable from an environmental point of view, because of the new externalities created, as we shall see.

There is a distinct pattern emerging here in Norway and in France, and to a degree in Japan, exemplified in a wider 2022 study of vehicle taxes covering 18 states from Iran to Malaysia. It included several countries such as Japan and the Philippines, which already have a weight tax [4.26].

There is undoubtedly a tension between weight-based and emissions-based approaches to vehicle taxation. Despite the trends toward heavier vehicles, the International Council on Clean Transportation (ICCT) in a recent report argued that countries should switch away from taxes based on vehicle weight and engine size and toward formulae based on emissions.

> Most countries, especially those with high taxation levels, would benefit from changing the current displacement- or weight-based taxes to continuous fuel efficiency/carbon dioxide-based tax based on linear metrics. Taxes based on engine displacement or weight incentivise the purchase of lighter vehicles and vehicles with smaller engines, but these parameters are not necessarily associated with better environmental performance of the vehicle [4.26].

Its argument rests on tailpipe emissions being a better indicator of overall environmental effect than vehicle weight or engine size.

In later chapters, we will examine which is the better approach. What is true to say now is that tax systems around the world at the moment lean more toward this emissions-based approach, but with some hallmarks of the weight-based approach. This, in part, accounts for the complexity of most systems. They are trying to accommodate the merits of each approach. What we can say at this point is that any optimal taxation system reflects the reality of the underlying market, society, and set of products you are taxing. What is optimal for a world of primarily ICEVs may not be best for a world of electrification. The ICCT might, therefore, be right for the past, but possibly not the future. The Tesla Cybertruck (2995 kg) is perhaps the best mental test: Should that be taxed high or low? Is it the ultimate green, zero-emission vehicle, or a tank of environmentally destructive heft?

It is a perhaps welcome sign of our heavier times that the European lobbying group for the automotive industry, the European Automobile Manufacturers' Association (ACEA), late in 2023 called for a kei-class classification to be introduced formally in Europe, to encourage small, light, urban electric vehicle solutions [4.47].

Despite these initiatives, weight-based taxation remains the exception, and it is often undermined by exemptions for favored technologies, even if the vehicles with these technologies are much heavier. Yet, current taxation approaches are coming under increasing pressure from declining revenues, the market realities of what vehicle manufacturers are producing, and better understanding of the environmental effects of heavier vehicles, whether they contain ICEs or are battery electric.

5

A Thought Experiment

IN THIS CHAPTER

- Consumer confidence in information from the automotive industry was shaken by Dieselgate.

- Technology options are getting more diverse and complex, including hybrids, electric vehicles, e-fuels and even hydrogen cars.

- If you could know only *one* piece of information to estimate the total environmental impact of your vehicle, and to compare that impact to other vehicles, what would it be?

So far, we have seen that cars are getting fat, people want to do the right thing for the environment, but everything is very complicated and confusing. One result of this is vitriol on internet forums. In a complex world, people cling to simple positions and then seek to rubbish different views, as tackling that complexity is too difficult. But if that simple position is not based on facts or the real world, it promotes an unrealistic, suboptimal, or even damaging solution.

The Problem

Vehicles have existed for hundreds of years; in this, we include horse-drawn carriages, and we have been monitoring their environmental impact for decades. For a consumer, in practice, this means that they see information such as fuel economy, carbon dioxide emissions, and perhaps maximum noise in the vehicle brochure or online. Other than these, consumers are aware, at a high level, that vehicles being sold new are subject to regulation and certification and so "the government is taking care of it." In other words, the car buyer can, by and large, choose the most useful and value-for-money car for them overall, and trusts that the system of vehicle certification is making sure that no car does particularly grievous damage to the environment.

This confidence in vehicle certification was profoundly shaken by the Dieselgate nitrogen oxides emissions scandal in 2015, which was described in previous chapters [5.1]. Even though a vehicle was certified to be emitting, for example, less than 180 mg/km of nitrogen oxides, the reality averaged about five times that [5.2]. As certification values were then questioned more and more, a focus also fell on carbon dioxide emissions, which were often a quarter above the values in the brochure, and fuel economy values were similarly below what was promised [5.3]. Therefore, events from 2015 onward have created an environment of distrust in official figures and certification systems; many consumers have started asking more questions and requesting more data, so that they can make an informed choice. This led to the launch, among other systems, of the AIR Index and Green NCAP—these were not official systems, but real-world data provided by reputable private organizations [5.4, 5.5].

But this was not the end of the story, because the market was about to become much more complex. As governments started penalizing ICEVs and incentivizing electrification, to try and draw a line under Dieselgate, this introduced another new technology, with different characteristics and performance—the BEV. Different technologies have coexisted before, most notably gasoline and diesel engines, with smaller contributions from natural gas and various hybrids. However, gasoline and diesel engines share many similarities—they burn liquid fuels at certain rates and in doing so produce a range of pollutant emissions that are released into the air. These vehicles are of similar designs and weights. In contrast, BEVs are rather different: no liquid fuel is burnt, no combustion pollutants are released, and the vehicles are quieter, but are heavier [5.1]. As a result, from the consumer point of view, the job of choosing a new vehicle just became much more complicated, as these different technologies must be compared in a fair way.

In other words, today, the world is changing, and the vehicle fleet is diversifying. No longer is every car powered by an ICE fueled with a fossil fuel. Some cars use lower-carbon fuels such as biofuels, and others use batteries and electric motors. As we progress through this century, this diversification is almost certainly going to increase. We may, ultimately, end up with another more homogeneous vehicle fleet again, but to reach that point will take decades, and it may never happen.

Although BEVs are gaining the most attention now, there are many other diverse technologies either in the market or coming. The plug-in hybrid vehicle, for example, is a particularly difficult technology to rate and compare to other cars [5.1]. These are vehicles that have both an ICE and an electric motor, which are used to different degrees during different modes of driving. The aim is to get the best of the efficiency of the battery motor at low speeds and the ICE at higher speeds, as well as the higher energy density of gasoline for longer range. For these cars, the tailpipe emissions, fuel economy, and noise are consequently highly dependent on the driving routes and styles of the individual driver. Furthermore, how often the owner charges it up also has a big effect. Unlike a pure BEV, these plug-in hybrids do not ever have to be charged up. While you might say that defeats the point of having them, there are in fact some perverse tax incentives with company cars in some countries that encourage owners not to charge them up [5.6, 5.7]. How can performance information and ratings statistics incorporate all these elements in a usable way?

Biofuels and e-fuels add further complexity. Ethanol has been blended into gasoline almost since the beginning of the ICE, originally in response to oil shortages during wars in the twentieth century [5.8]. In the US, this started to happen on a more systematic basis in the 1970s, motivated by energy security during the oil crises [5.9]. Ethanol could be produced from crops grown locally and mixed in various proportions into gasoline to reduce fossil fuel use without modification, or at least significant modification, to the engine [5.1]. When burnt in the engine, the emissions are broadly the same as standard gasoline, although with some differences, such as aldehyde emissions that arise from the "bio" compounds that contain oxygen [5.10, 5.11]. The principal emissions benefit from biofuels is that the crops absorb carbon dioxide when grown, resulting in lower emissions on a lifecycle basis, as we will see in more detail in a later chapter. But how does a consumer know the amount of ethanol in their gasoline, where that ethanol has come from, and how it was grown? Before you say that we know that "E10" gasoline you buy at the pumps contains 10% ethanol, it does not. "E10" means *up to* 10% ethanol. Empirically, it tends to be close to 10%, but could legally be zero. A similar situation exists for diesel, where other "bio" sources are used such as waste cooking oil in the "B7" diesel you buy at the pump in Europe and the US. What this all means is more complexity for consumers in trying to understand emissions and do the right thing.

Even though the excitement about the transition to BEVs is the loudest, the reality in the market is that more and more biocomponents are being blended into liquid fuels around the world, and this trend is only likely to continue. The limiting case of this, one could argue, is e-fuels. These are made through synthetic chemistry by combining carbon dioxide with hydrogen to create non-fossil hydrocarbons by different methods that can replace gasoline and diesel. These fuels have the potential to be near-zero for carbon dioxide emissions [5.12]. How would we know if they were blended into existing gasoline and diesel, displacing the fossil fuel component in the same way as with biofuels? This would present yet another complexity that the consumer would be ill-placed, and quite likely disinclined, to fully understand.

In these few brief paragraphs, we have shown the multitude of technologies that are available in cars, covering the motor, its fuels, and how they interact with driver behavior. Complex and getting progressively more complex. That said, it is possible to envisage a world where the car fleet is largely made of pure BEVs. That is certainly the policy of many Western governments for 2050, to meet various net zero goals [5.1]. In that world, the job of the consumer from an environmental point of view becomes much easier again. No tailpipe emissions. No blended fuels. No unpredictable hybrid systems. Just one or two electric motors powered by known grid electricity.

However, that ultimate outcome is here not certain. In any case, we are in a transition, even if we do not know to what we are transitioning. Therefore, consumers, governments, and the wider market need help today to assess this fiendishly complex set of technologies in such a way that good decisions can be made. Without good information, how can we expect to make good decisions?

The Thought Experiment

Therefore, we need something that is simple enough and accurate enough. So, we asked ourselves this question:

If you could know only *one* piece of information to estimate the total environmental impact of your vehicle, and to compare that impact to other vehicles, what would it be?

Is That Even Possible?

Usually, to develop ratings of products, one seeks to capture all the relevant factors into some beautifully combined index. This is what Green NCAP is [5.13]. Beautiful and complex. But to be successful, useful consumer information needs to be simple, intuitive, and accurate enough. The solution to our reservations with Green NCAP is absolutely *not* to add in more factors, but to consider whether many of the factors can be excised. A good case for Occam's razor.

In posing the question, we must accept that the answer might be that there is no single piece of information that is accurate enough. Clearly, the environmental impact of cars has many and varied dimensions, and we are not naïve enough to expect that they will automatically and easily cancel down to something obvious. You would be right to ask whether there is even a reasonable chance of success, given all the aspects from greenhouse gas emissions, air pollutants, and noise pollution, from production to disposal of the vehicle, and wherever the vehicle is used and in what way it is used. You could even be forgiven for thinking it is an impossible task. But read on.

To put the task in perspective, we would also venture to say that it is *impossible* to create some index or ratings system that incorporates *every* aspect in which a vehicle impacts the environment. Pollutant emissions, greenhouse gas effects, noise, and so on have a multitude of dimensions and some are subjective, so the perfect index is a myth. In addition, how do you weigh such dimensions? Which do you consider to be more important? Coming up with objective weightings would be very difficult. No system can be expected to reflect the features of every vehicle and every person's behavior. By adding more and more factors, we give ourselves a false sense of accuracy, at disproportionate effort. The marginal extra complexity and testing cost of adding in a variable of tertiary importance are high.

We must also consider what constitutes "accurate enough." Perhaps if half the environmental impact could be accounted for by one factor, in a world where there are dozens of factors at play, that may count as success. The prize, if successful, would be significant—consumers would be able to make easier, more accurate decisions, while doing the right thing for the environment.

To address this question, we racked our brains and scoured our databases of real-world emissions test data, and came to the conclusion that such a singular parameter did exist.

And this leads us to propose the *Molden-Leach Conjecture.*

The following chapters of this book will set out the rationale that such a parameter does exist, test its suitability, and explain what it is.

6

The Proof

IN THIS CHAPTER

- There are many pollutant measures that could be that "one piece of information" sought in Chapter 5.

- Each pollutant varies with the design of the car and how it is made and used.

- The only parameter that tracks environmental impact well is vehicle weight: the *Molden-Leach Conjecture*.

As we progress toward the *Molden-Leach Conjecture*, now is the right moment to answer the question we have just posed:

If you could know only *one* piece of information to estimate the total environmental impact of your vehicle, and to compare that impact to other vehicles, what would it be?

As we unpack the question, it is worth taking a step back immediately. Is it in fact possible to determine the total environmental impact at all? We would argue that it is almost impossible to determine the total environmental impact of a vehicle *exactly*. There is so much complexity in this system and dependence on the behavior of people that it is incredibly unlikely that we can obtain all the information needed over the lifetime of a vehicle from its inception to its final resting place. Clearly, if we cannot get an exact assessment of the total environmental impact of a vehicle from knowing absolutely everything about it, then doing so from one piece of information is clearly impossible.

So, should we even bother?

For a long time, we were relatively good at estimating the environmental impact of a wide array of humankind's activities. When there is enough of something around, averages and estimates can become very useful. Indeed, almost all the taxation we looked at in Chapter 4 relies on averages—whether that is average fuel consumption or average carbon dioxide emissions. We clearly do not know exactly how everyone drives, or how long you spend sitting in traffic, or exactly which country the barrel of oil that produced your gasoline came from, or which power station produced the electricity that you charged your BEV with. Yet, we do know that when we take all the vehicles in a given situation, for example vehicles of the same type or in the same location, we can have a good idea of their environmental impact. As noted in Chapter 4, depending on where you are in the world, the information typically used today to determine a vehicle's tax burden, which is often aimed at accounting for the total environmental impact of your vehicle, includes some combination of fuel consumption, fuel type, tailpipe carbon dioxide emissions, sticker price, power, and weight. In a world dominated by ICEVs, we get a good idea of a vehicle's total environmental impact from these six parameters.

We have, nevertheless, entered a trade-off between simplicity and accuracy. We are never going to know the impact exactly, so we are using six variables to get a good estimate of a vehicle's total environmental impact. But why six? Is that the best mix of simplicity and accuracy? If we can get similar accuracy with fewer variables, that makes a consumer's car purchasing decision that much easier. So, how about just one? We will argue that one can work, and we are going to spend the rest of this book explaining exactly how.

What Should That Piece of Information Be?

To test the *Molden-Leach Conjecture*, then, what might we choose? An obvious place to start is the parameters listed above: fuel consumption, fuel type, tailpipe carbon dioxide emissions, sticker price, power, and weight.

On fuel consumption, not all vehicles are using a fuel, in the classic sense of the word, anymore.[1] If we changed this to energy consumption, we might have a candidate. With regard to tailpipe carbon dioxide emissions, not all vehicles emit carbon dioxide *from the vehicle* anymore (BEVs and FCEVs do not) [6.1]; we cannot take into account the impact of sustainable fuels either as they emit carbon dioxide from the tailpipe, which we can measure, but this does not take into account the absorbed carbon dioxide in their production. This metric, therefore, must be discarded at an early stage. Power and mass are possibilities; we should keep them in play.

Other obvious choices include the following: driving range, vehicle size, powertrain (ICE, hybrid, battery electric, hydrogen, and so on), powertrain size (battery or engine capacity), torque delivered to the wheels, top speed, monetary value, geographical location of vehicle use, and geographical location of vehicle production. There are, no doubt, more than you can think of.

Having considered what parameters might be important, we should also think about how to evaluate the environmental effect of a vehicle. When we consider the environmental impact of a vehicle, it is going to be important to consider the impact over the whole life cycle of a vehicle on the environment, not just the impact of driving it around, like we do today. To do that formally, we need to undertake a life cycle assessment (LCA)—and we take a look at this in more detail later in the book. However, for now, let us just make sure we consider all the life cycle of the vehicle in the *Molden-Leach Conjecture*. A useful breakdown is shown in Figure 6.1, where we consider the raw materials for manufacture, their processing, the production of the vehicle, the production of

---

[1]   What is a fuel is an interesting question. Most people use the words in the sense of being a primary energy source—energy found in nature that has not been subject to a human-designed conversion process. Electricity is not a fuel by this definition, as it is derived from other primary sources (the sun, wind, coal, oil, natural gas, uranium, to name but a few). Fossil-derived gasoline and diesel are fuels, as crude oil is their source and the energy in gasoline and diesel is overwhelmingly from that crude oil—it has not needed to be converted. Hydrogen, biofuels, and e-fuels are not fuels, as their contained energy is converted from other primary energy sources.

the "fuel" (electricity, gasoline, whatever), the use of the vehicle, and, finally, its disposal and recycling. It will be convenient for this thought experiment to split these into two parts: the "fuel and in-use" and the "raw materials, processing, production, and disposal."

FIGURE 6.1 The phases considered in LCA.

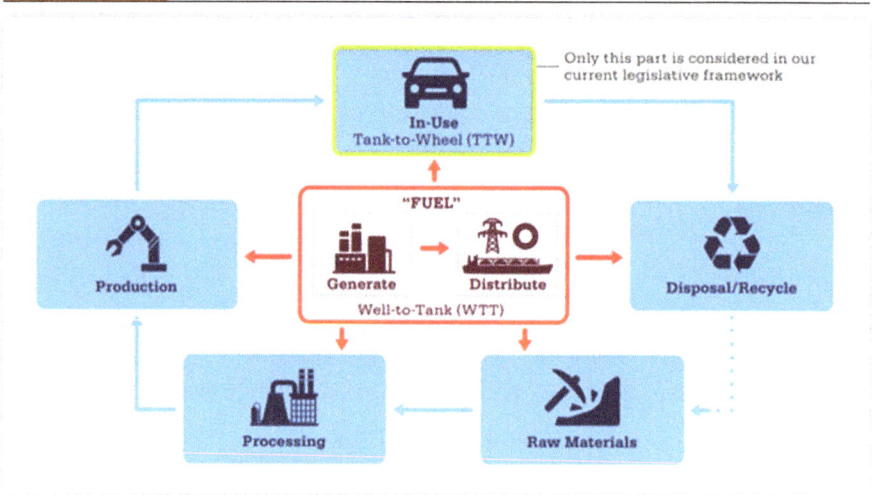

So, now we have a list of candidates and some LCA "bins" against which to assess their efficacy as metrics of the total environmental impact of a vehicle. The resulting candidates are:

- Energy consumption
- Power
- Mass
- Range
- Physical size
- Powertrain (ICE, hybrid, battery electric, hydrogen)
- Powertrain size (engine capacity, battery size)
- Torque at the wheels

- Top speed
- Cost new
- Value today
- Geographical location of use
- Geographical location of production

LCA: Fuel and In-use

To begin to think about these phases, let us go back to first principles about a vehicle and the forces that act on it. We can then consider the forces acting on the vehicle to be: (1) the propulsive force driving it forward, and this being counteracted by (2) air resistance, (3) friction with the road, also known as rolling resistance, (4) weight (only when going uphill or downhill), and (5) internal mechanical losses, or friction between the moving parts of the engine and powertrain. There may be some other small effects too, such as disturbance forces, for example from driving over a pothole, but for the purposes of this exercise, we will neglect them. The forces acting on our vehicle are shown in Figure 6.2, and let us consider each of them in turn.

FIGURE 6.2 Forces acting on a vehicle in motion—in this case driving uphill.

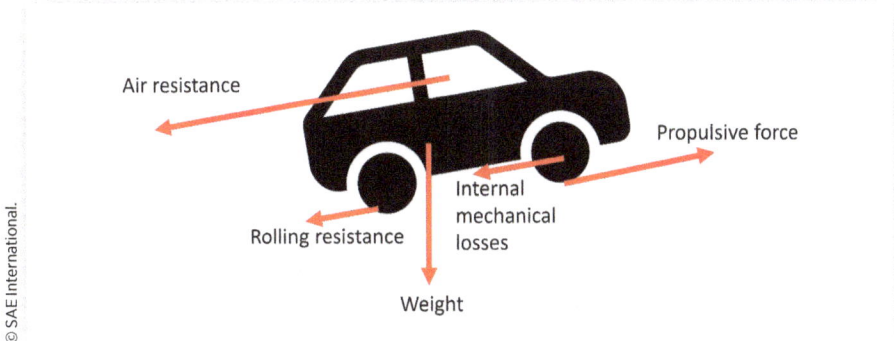

Air resistance

Propulsive force

Internal mechanical losses

Rolling resistance

Weight

© SAE International.

Air Resistance

The air resistance (aerodynamic drag force) depends on the air density (ρ_a), vehicle velocity (throughout this chapter, we will use the word velocity, the vector form of speed), the frontal area of the vehicle (A_f), and effectively how aerodynamic it is (captured by a drag coefficient—C_d). These are combined to make the following equation for the drag force because of air resistance:

$$F_{drag} = \frac{1}{2} \times \rho_a \times A_f \times C_d \times v^2$$

Values of C_d will vary depending on the shape of the vehicle within the range approximately 0.2–0.5, with higher values indicating higher drag. Values of C_d for a range of vehicles of different shapes and sizes are shown in Table 6.1. You will observe that the range of C_d for what most people would consider "normal" cars is 0.2–0.35, but significant advances in aerodynamics have been made recently and vehicles from the 2010s onward are often showing C_d values at the bottom of that range, certainly below 0.25.

TABLE 6.1 Values of C_d for a variety of production vehicles.

C_d	Model year	Vehicle	Source
0.58	1997	Jeep Wrangler (softtop)	[6.2]
0.52	1984	Jeep Cherokee	[6.2]
0.52	2002	RAM 1500	[6.2]
0.48	1938	VW Beetle	[6.2]
0.40	2000	Chrysler PT Cruiser	[6.2]
0.36	1992	Opel (Vauxhall) Corsa	[6.2]
0.35	1992	Nissan Micra	[6.2]
0.34	1993	Chevrolet Camaro	[6.2]
0.33	2001	Mini Cooper S	[6.2]
0.31	2012	Pagani Huayra	[6.3]
0.31	1998	Toyota Corolla	[6.2]
0.25	2010	Toyota Prius	[6.2]
0.23	2017	Tesla Model 3	[6.4]
0.22	2019	Porsche Taycan	[6.4]
0.22	2018	Mercedes A-class	[6.4]
0.22	2011	BMW 5 series	[6.4]
0.21	2012	Tesla Model S	[6.4]
0.20	2021	Mercedes EQS	[6.4]
0.20	2022	Lucid Air	[6.5]
0.20	2024	Xiaomi SU7	[6.6]
0.19	1996	General Motors EV1	[6.2]

It is possible to get much lower values of C_d—for example, vehicles that race in the Shell Eco-marathon can have C_d values as low as 0.045 [6.7]. However, the look and practicality of these vehicles (see Figure 6.3) rule out values this low for practical applications.

FIGURE 6.3 Vehicles racing in the Shell Eco-marathon. Would you own a car like this?

Abdul_Shakoor/Shutterstock.com.

Friction (Rolling Resistance)

The friction, usually termed the rolling resistance, can be determined as a coefficient representing the tire–surface interface (C_R)—this represents how well inflated your tires are, their tread and type, what type of surface you are driving on, and so on—multiplied by the normal contact force (R) between the surface and the vehicle. You may remember $F = \mu \times R$ from your school mathematics or physics class, where μ is the coefficient of static friction. Now we are moving, we need to modify this a bit.

More formally we can express it as:

$$F_{rolling} = C_R \times m \times g \times \cos(\theta)$$

where C_R is the rolling resistance coefficient (dimensionless), m is the mass of the vehicle, and g is the gravitational acceleration

($m \times g$ is the normal contact force when the vehicle is on a flat surface; for a hill, multiply by the cosine of the angle of the hill—let us call this angle "theta" (θ)). C_R will vary depending on what road surface you are driving on and, to an extent, your tires. Just so you have an idea, typical values of C_R are shown in Table 6.2 (again, higher values indicate more resistance).

TABLE 6.2 Values of C_R for a variety of driving surfaces assuming standard tires [6.8].

C_R	Road surface
0.01–0.015	Concrete; new asphalt; small, new cobbles
0.02	Tar; asphalt
0.02	Rolled new gravel
0.03	Large, worn cobbles
0.04–0.08	Solid sand; loose or worn gravel
0.2–0.4	Loose sand

© SAE International.

Weight

When traveling uphill, the propulsive force will also need to overcome a contribution from the weight of the vehicle (see Figure 6.2 again). This will be related to the angle of the hill that the vehicle is driving up—θ. In fact, when you work through the mathematics, the weight contribution is proportional to the sine (the trigonometric function that gives us the contribution in a given direction) of the angle that the vehicle is driving up.

$$F_{weight} = m \times g \times \sin(\theta)$$

When traveling downhill, the reverse is true. However, the impacts of this now depend on your vehicle powertrain. In a vehicle fitted with regenerative braking, such as a hybrid or BEV, you will be able to recapture much of the energy to charge your battery. In a vehicle that is not fitted with regenerative braking, such as most conventional ICEVs, you cannot capture this energy, although you may be using the engine less travelling downhill, and this energy is lost to heat mainly through the brakes.

Internal Mechanical Losses

Internal mechanical losses are complicated to calculate exactly, but will be proportional to the velocity at which the components rotate, as most of these losses are frictional. If we assume that all the relevant components rotating are lubricated, then much of the losses will be because of shear stress (effectively the retarding force caused by fluid friction) in the lubricating oil (τ). Assuming that the lubricating oils are Newtonian fluids (ultimately most "normal" fluids are, which means their viscosity does not change with the applied force), then this shear stress will be expressed as:

$$\tau = \mu \frac{du}{dy}$$

In this equation, μ is the viscosity of the lubricating oil, that is, how sticky it is—honey has a relatively high viscosity and water has a relatively low viscosity, for example. $\frac{du}{dy}$ is the velocity gradient (difference in velocity per unit distance) between two moving surfaces; given that almost all the relevant components in a car are rotating, the velocity gradients between these rotating parts will almost all be proportional to the velocity that the vehicle is traveling at.

This means that we can express the mechanical losses as proportional to the velocity of the vehicle and the normal contact force of the interface between those two moving surfaces, just like the friction equation above.

$$F_{mechanical} = k \times v \times m \times g$$

k in this equation is a constant that takes into account both the oil viscosity and some coefficient of friction proportional to the normal contact force. Note here that the mass (m) for an individual interface between two surfaces will not in all cases be the mass of the whole vehicle—for example, in a car with two axles, the bearings for each axle will only have half of the mass of the car acting on them, but when we sum the mechanical losses over the whole vehicle, the mass of the vehicle will be a good approximation. In other words, every part of the vehicle mass will be transmitted through a rotating joint somewhere, most notably in the axles.

Propulsive Force

The propulsive force driving the car forward, then, will need to be equal to or greater than the sum of these other forces, if the vehicle is to actually move. If it is equal to the other forces, the car is moving at a constant velocity on the flat, or stationary. If it is greater than the sum of the other forces, then the car is accelerating or going uphill. If the car were on an incline, the additional force required would be simply to overcome gravity and so directly proportional to mass (mg). Therefore, we can write an expression for the propulsive force as follows:

$$F_{propulsive} \geq F_{drag} + F_{rolling} + F_{weight} + F_{mechanical}$$

When expanded it becomes:

$$F_{propulsive} \geq \left(\frac{1}{2}\rho_a A_f C_d v^2\right) + \left(C_R m \cos(\theta)\right) + \left(mg\sin(\theta)\right) + \left(kvmg\right)$$

So, where does this leave us? Creating the propulsive force is where much of the environmental impact occurs, through using fuel. This is what fossil fuel, sustainable fuel, hydrogen, or electricity is being used to provide, and the more we drive, the more energy to provide this propulsive force we need. So, if we can understand the things that affect the propulsive force, we can begin to see where we can influence our environmental impact. As an aside, some other environmental impacts such as non-exhaust emissions, which we will look at in a later chapter, also come from the $F_{rolling}$ and F_{weight} contributions because they—in particular, tire wear emissions—depend on the friction force between the vehicle and the road.

However, there are many things in this propulsive force equation that we can do very little about, which are summarized in Table 6.3.

TABLE 6.3 Ability to change parameters making up propulsive force.

Parameter	Can we change this to affect our propulsive force requirement?
ρ_a	This is the air density. You cannot change this.
A_f	The frontal area of the car. We can affect this, but the ability to reduce it is constrained by the utility requirement of the vehicle.
C_d	We can reduce this coefficient of drag by improving the vehicle aerodynamics, but for a given size vehicle, and user desires about vehicle aesthetics, it is rather constrained.
m	There are, of course, constraints on this, but masses of vehicles that users drive vary substantially—a factor of 3.5 difference is possible.
g	Provided you are driving on earth, you cannot change this.
k	Using better lubrication oils can help with this, and much progress has been made in this area in recent years, but there are limits.
C_r	While variations in rolling resistance are possible both down to tire design and inflation pressure [6.9], there are limits as to how low this can go.
θ	This is how steep the hill is. To some extent, it is the user's choice which hills we drive up and down.
v	We can reduce the velocity we drive at, but in general this is down to user choice.

We can group these things into items we cannot change at all (ρ_a, g, θ), items that are hard to change by much in practice or users are resistant to change (A_f, C_d, v, C_r, k), and mass (m).

With respect to our candidate parameters that we discussed previously, it is clear that from our first principles analysis, mass is a leading candidate, but we should consider the others too. This is summarized in Table 6.4.

TABLE 6.4 Influence of candidate parameters on in-use environmental impact.

Parameter	Influence on in-use environmental impact
Vehicle energy consumption	Will be well correlated—more energy use will lead to higher environmental impact.
Vehicle power	More power may be linked with higher environmental impact, but there will be a big influence of powertrain and fuel type. There is some evidence that higher vehicle power is linked to more aggressive driving, which will have a higher environmental impact [6.10].
Vehicle mass	Closely correlated when looked at from first principles (see the earlier analysis).
Vehicle range	Unlikely to be well correlated, except indirectly for a BEV, as greater range often corresponds to greater mass.
Vehicle size	Will be well correlated when considering A_f, for example. Will also be correlated with mass, so indirectly through that route as well.
Vehicle powertrain	Highly correlated but only between categories—that is, comparing battery electric vs ICE vs hybrid vehicles, and so on. This is a coarse metric.
Vehicle powertrain size	Likely to be slightly correlated—bigger ICEs or electric motors will use more energy.
Torque delivered to the wheels	Torque (rotational force) is likely to be well correlated—as it will be closely correlated with vehicle mass and vehicle energy consumption. There is some evidence that higher vehicle torque is linked to more aggressive driving, which will have a higher environmental impact [6.10].
Top speed of vehicle	Not likely to be well correlated as a vehicle rarely travels at its top speed when in use.
Cost of vehicle when new	Unlikely to be correlated.
Value of vehicle today	Unlikely to be correlated.
Geographical location of vehicle use	Unlikely to be well correlated, but other factors such as local grid intensity will affect BEV and plug-in hybrid in-use emissions.
Geographical location of vehicle production	Will not be correlated at all.

© SAE International.

So, considering this part of the analysis, the likely candidates for our single parameter are energy consumption, vehicle mass, vehicle size, and torque delivered to the wheels.

LCA: Raw Materials, Processing, Production, and Disposal

Let us now consider the influence of these same candidate parameters on the other LCA "bin," the environmental impact of the raw materials, processing, production, and disposal. These are summarized in Table 6.5.

TABLE 6.5 Influence of candidate parameters on LCA aspects of environmental impact.

Parameter	Influence on raw materials, processing, production, and disposal environmental impact
Vehicle energy consumption	This may be indirectly correlated with vehicle mass, but it is easy to imagine scenarios where this might be poorly correlated with this "bin"—particularly when there are different powertrains, for example BEVs will have lower energy consumption but typically higher LCA emissions in this bin.
Vehicle power	More power usually correlates with bigger cars, but it is an indirect link and will break down for vehicles such as sports cars.
Vehicle mass	Closely correlated. The heavier your vehicle, the more materials you need to produce it.
Vehicle range	For a BEV, this is likely to be closely correlated; for an ICEV or hybrid, it is likely to be uncorrelated.
Vehicle size	Possibly correlated, but would break down for vehicles with high levels of empty space.
Vehicle powertrain	Will be correlated. Greater impact for a BEV, lesser for an ICEV.
Vehicle powertrain size	May be correlated for a BEV, but unlikely to be correlated for an ICEV.
Torque delivered to the wheels	Given the association with vehicle mass, may be closely correlated, but it is an indirect link.
Top speed of vehicle	This is unlikely to be well correlated as many mass-market vehicles have similar top speeds and sports cars will skew this significantly.
Cost of vehicle when new	It is possible this will be correlated; for example, if production is more expensive, then a vehicle will cost more, but there are many factors that feed into the cost that it is unlikely to be well correlated.
Value of vehicle today	Unlikely to be well correlated. Used car values are driven by many other market factors.
Geographical location of vehicle use	There will be a minor contribution as vehicles need to be transported from their production location to their usage location, and there will be emissions associated with this.
Geographical location of vehicle production	Will have some correlation as production emissions are dependent on local factors such as grid carbon dioxide intensity, but does not take into account vehicle-specific factors.

Picking a Winner

It is clear that there is one parameter that is closely correlated with life cycle environmental impact both from first principles and from an analysis of a wide range of likely candidate parameters.

Mass.

Nothing else comes close.

We have arrived at the *Molden-Leach Conjecture*. Vehicle mass is a single parameter that we can use to estimate fairly and usefully accurately the total environmental impact of a vehicle over its whole life cycle.

Mass has a wide-ranging impact, from requiring more energy and material for manufacture, to needing more energy during driving, to creating higher wear emissions from the tires and road, to higher noise, to... Well, you get the idea, we are going to be writing the rest of the book about this.

Given the title of this book, that we have ended up with mass will not surprise you, but I hope that you are becoming more convinced of the importance of this parameter above all others. We hope to demonstrate that we can make a good estimate of the total environmental impact of a vehicle knowing only that vehicle's mass.

Decoupling of Powertrain from Emissions

Notice that, so far, what is providing the propulsive force for our vehicle is completely immaterial. This could be an ICE through some gears, an electric motor, a fuel cell, or indeed anything else up to the job. All we need is vehicle mass. This is a substantial change to established policy in most countries where the diversification of the fleet has been reflected in an increasingly complicated labeling and taxation system. In California, for example, a BEV, which legally is a ZEV, can count as more than one vehicle (weird, right) for the purpose of credits toward a manufacturer's zero-emissions fleet. This is to incentivize supposedly ZEV uptake. Regardless of what you think of this, it is clearly not sustainable into the future and may lead to misleading environmental impact assessments. Our proposal avoids all these complications, is simple, and, we think, effective.

But That Is Not All

One important detail to this, however, mass on its own is not enough. Wait, what? Does not this undermine our whole thought experiment earlier? Inevitably, the environmental impact of a vehicle when it is being used is also determined by how *much* it is used. The thought experiment asked:

If you could know only *one* piece of information to estimate the total environmental impact of your vehicle, and to compare that impact to other vehicles, what would it be?

We now have that piece of information—the *Molden-Leach Conjecture*—which describes the total environmental impact across all pollutants *per unit of* travel activity. To calculate the aggregate impact of the environment we need to then combine that with activity information in the form of how much that vehicle is driven.

With this proposal, there is an interesting parallel to observe. The force required to move the vehicle is proportional to its mass and, when moved through a distance, "work" is "done." As any school physics student will be able to tell you, work done has units of energy (joules) and is determined by this equation:

$$\text{Work done} = \text{Force} \times \text{Distance}$$

So, the environmental impact of a vehicle is broadly proportional to the work done in moving it around!

In almost all jurisdictions, vehicles have to report their annual distance traveled (mileage) at some point, whether for renewal of registration or at some annual technical inspection. We, therefore, have good data on distance that are regularly updated. As a result, we propose that we can, to a decent level of accuracy, deduce the total environmental impact of a vehicle through the *Molden-Leach Conjecture* very simply:

$$\text{Environmental impact} = \text{Vehicle mass} \times \text{Annual distance traveled}$$

We will explore this proposal in detail throughout the rest of the book.

7

Tailpipe Pollutants

IN THIS CHAPTER

- Fossil-fueled ICEVs are a major source of emissions.
- Real-world data shows that car weight is a good predictor of tailpipe carbon dioxide emissions.
- Air pollutant emissions are broadly uncorrelated with weight, but have mostly been effectively abated already.

Now we have our proposition that the best single measure of a vehicle's environmental effect is its mass, we must test this against real-world emissions data. Thinking from first principles, we have seen that it is reasonable to believe that mass is a good metric—the heavier the vehicle, the more fuel it should need to move along, other things being equal. More fuel means emissions from the tailpipe rising in proportion. So, perhaps this will prove a simple and uncontroversial hypothesis? Fat chance. Nothing in emissions is simple. But that is why we need more than ever a simple proxy that is representative of reality to work with.

How good is the correlation between mass and emissions in all their forms? Are there better proxies than mass? Of course, no single variable can be a perfect predictor, far from it. However, as discussed in previous chapters, we are looking for the sweet spot, the Goldilocks zone, between simplicity and accuracy, so consumers and producers can make decisions that align as best as possible with environmental goals, without being so complicated they get ignored. It is quite likely that the correlation will vary between different pollutants, which will leave us to form a view in the end whether mass, overall, correlates well enough with the most damaging pollutants.

Are Emissions Only from the Tailpipe?

The range of pollutants to consider is many and diverse. When thinking of vehicles and the environment, the mind tends immediately and inevitably to the tailpipe, the protruding conduit that neatly ejects all the sins of combustion in the engine. But this forms only part of the picture [7.1]. Even within that exhaust pipe, there are pollutants that are familiar and some less so [7.2]. The familiarity tends to follow from a pollutant being regulated by some authority, which tends to spawn a body of literature, consumer information, and media coverage. The less familiar pollutants are those that can broadly be thought of as the unregulated ones. While we may be relatively familiar with nitrogen oxides thanks to the Dieselgate scandal, acetaldehyde—despite being a gas internationally classed as "probably carcinogenic" and associated with organ damage in humans—is much less well known as it is regulated almost nowhere in the world, except to a limited extent in Brazil [7.3, 7.4].

Standing back from the tailpipe, vehicles emit many substances from other parts. Whether in operation or stationary, small amounts of fuel will be evaporating from the tank of a ICEV [7.5]. On modern cars, there is technology installed to capture and return these fugitive gases to the tank, but they can never be perfect. From exterior plastics of a car, including tires, volatile organic compounds, many of which can cause negative health effects and wider environmental damage, evaporate constantly, faster during warm days and slower during colder nights [7.6].

In motion, the same tires that off-gas the volatile organics wear down and, in the process, release rubber particles, especially as the car brakes and corners [7.7]. During that braking, the brake pads and discs, or in some cases the older-style drums, wear too, releasing a range of tiny particles. Symmetrical with the wear of the tires is the abrasion of the road [7.8]. The force between the tires and road shears off small pieces of asphalt or concrete. Thinking even wider, one could include the volatile organics always off-gassing from that asphalt laid down to allow the car its smooth path [7.9], and the random material that has settled on the road from neighboring fields for example, which is stirred up as the car passes—often called "resuspended" particles [7.10]. Together, this group of pollutants being emitted from parts of the car other than the tailpipe are usually called "non-exhaust" pollutants.

The exhaust and non-exhaust emissions are collectively seen as the "in-use" emissions of the vehicle, although this is perhaps a slightly misleading term as some emissions happen all the time, whether the car is in use or not, such as the fuel evaporation [7.11]. Prior to the use phase of the car, it has to be manufactured, which creates a range of emissions itself. Mining the raw materials and the transportation, processing, assembly, and distribution of the completed vehicle all incur carbon dioxide emissions from the energy used, as well as air pollutants from this fuel combustion and non-air pollutants [7.2]. These non-air emissions could be liquid waste dumped into rivers—often legally—or solids. The manufacture of batteries is an especially toxic process, as excavated rocks tend to require crushing and washing in acid repeatedly as part of the refinement process. The volume of liquid waste is significant and growing fast as the BEV sector expands. In China, much of this is released into rivers, often quite legally because of environmental regulations much laxer than Europe or the US would tolerate [7.12]. When thinking about the manufacturing emissions of a vehicle, it is easy to miss an obvious one: the fuel used to power it. Tailpipe emissions are obviously a product of combusting the fuel in the car, which releases the embedded carbon dioxide and creates air pollution. But this fuel does not occur naturally and is the result of a process of refining crude oil. The drilling or extraction of this oil is energy-intensive and polluting, as is the process of cracking the crude into its chemical fractions, from which come gasoline and diesel, among many other products.

The final group of emissions are often not immediately thought of as emissions at all. Noise, for example. At low speeds, the main source of noise from a car is the moving parts inside the engine: the pistons, crankshaft, gearbox, and so on. Beyond a certain speed, tires take over as the dominant source of noise, as they are squashed against the road surface and air is expelled from tread voids [7.13]. These noises are a largely unwanted by-product of converting energy stored in the fuel to motion, even though the sound produced does have a few compensatory advantages, including warning pedestrians of an approaching vehicle. Nevertheless, it is generally considered that these sounds form "noise pollution," another negative environmental effect the driver does not suffer any cost or downside from causing.

When a vehicle crashes, it also creates a negative externality. A wall demolished or—much worse—a person injured or their car damaged are externalities. This external harm is being caused as a result of the driver's actions that he does not suffer a direct cost of. The insurance system, in effect, services as a way to internalize this externality, at least to some extent. If culpable, the driver will suffer a financial loss in terms of an insurance excess and potentially higher future premia. Nevertheless, there are many instances where this does not work, from chewing up grass verges to navigating a tight spot through to the impossibility of fully compensating for a third-party death caused. Heavy vehicles, in particular, such as trucks, often cause damage to the transport infrastructure, as evidenced by tire grooves pressed into asphalt on the inside lane of highways with high volumes of truck traffic.

Finally, they might also increase the aesthetic damage. Rows of parked cars in the center of a bucolic country town are probably not most people's idea of beauty. Their presence uglifies the place and reduces the utility and happiness of the residents. That is not to say on balance that these cars are not bringing benefits—in business and mobility—but there is an aesthetic externality that is being caused that the car owner is not paying for.

In short, therefore, we must avoid a narrow concept of vehicle emissions as only coming from the tailpipe, but entertain the in-use non-exhaust emissions, the construction emissions, and the other environmental bads. This chapter will look in detail at the first category—the tailpipe emissions—and subsequent chapters will tackle the others. So, let us return to the classic tailpipe emissions and see how they correlate with vehicle weight. For now, let us set aside all BEVs as they have no tailpipe emissions by design.

Official Certification Data versus Independent Real-World Data

All vehicles with an ICE are certified in their country before they can be sold legally. In the UK and Europe today, the certification test for carbon dioxide follows the Worldwide Harmonised Light Vehicle Test Procedure (WLTP), while carbon monoxide, nitrogen oxides, and particle number (PN) follow a combination of WLTP, which is conducted in the laboratory, and "real driving emissions" (RDE), which is an on-road test that checks that this value is close enough to the laboratory result [7.14]. Each car needs to pass both tests for every pollutant. However, WLTP and RDE were only introduced around 2018 [7.15]. The date of introduction is not precise as each test had slightly different, staggered introduction dates, and different dates for cars and vans. Prior to 2018, the certification was based solely on a simple laboratory test called the New European Driving Cycle (NEDC) [7.16, 7.17]. This is relevant because there is a world of difference between the two regimes. The NEDC, before 2018, was a very easy test, conducted only in the laboratory, and the rules were drawn with some nasty gray areas in the drafting. As a result, the emissions levels were very flattering, and vehicle manufacturers were able to apply all sorts of cleverness to exploit the legal gray areas to get the lowest emissions results. Some may say this was "cheating," but these knotty questions remain in courts across Europe in 2024. After 2018—in response to Dieselgate—the combined WLTP and RDE regime closed many (but not all—see Dornoff et al. [7.18]) of the loopholes and made the tests more challenging. As an aside, and for completeness, the US avoided similar issues by proactively tightening its laboratory test in 2008 by introducing its "five-cycle method" [7.19].

This piece of history is relevant because it means that official certification data is not comparable over time and, especially in respect of the previous data, bear little relation to real-world emissions, which are generated with real weather, roads, gradients, traffic, and so on. Therefore, it would be unwise to rely on this data to assess the effect of vehicle weight on emissions. Instead, we can consider the independent real-world data gathered since 2011 by Emissions Analytics (which was founded and is run by Nick Molden), using regulatory-grade measurement equipment [7.20]. In total, 1190 European vehicles have been tested from model years 1998 to date. These are a mixture of gasoline, diesel, and hybrid vehicles of different flavors from mild to plug-in (a detailed look at the differences between these types is given in Felix's previous book, with Kelly Senecal, *Racing Toward Zero* [7.2]). Each of the vehicles was tested for around three hours in a set mix of urban, rural, and motorway driving. To achieve representative driving, dynamic criteria were applied to ensure urban driving was sufficiently "stop–start" and included some congestion, and that motorway driving was fast enough. Extreme weather conditions were avoided, and driving styles were kept as consistent as possible, while using human drivers.

Pooling all the results together, whatever the type of fuel, we get the results shown in Figure 7.1 for carbon dioxide, the well-known "greenhouse gas" and a major cause of climate change. The vertical axis is the real-world emissions over the complete test cycle. The five different bars represent different "mass classes"—"$500 \leq m <$ 1000" means vehicles that weigh at least 500 kg (1100 lb), but less than 1 tonne (2200 lb). Each category covers a 500-kg range. The chart is a neat visual way of showing how the emissions vary within each category. Technically, the orange box describes the "interquartile range," which is everything except the 25% most extreme values at the low and high ends. It is the middle half of values. Within the orange box, the horizontal line represents the median (the middle value if you rank them all) and the "x" is the mean, the common notion of average. The lines stretching out of the orange box—called the "whiskers"—show the whole range of values except the top and bottom 5%. Finally, the dots are the "outliers"— the more extreme values that can skew the overall picture.

FIGURE 7.1 Carbon dioxide emissions for different ICEV mass classes for 1190 vehicles.

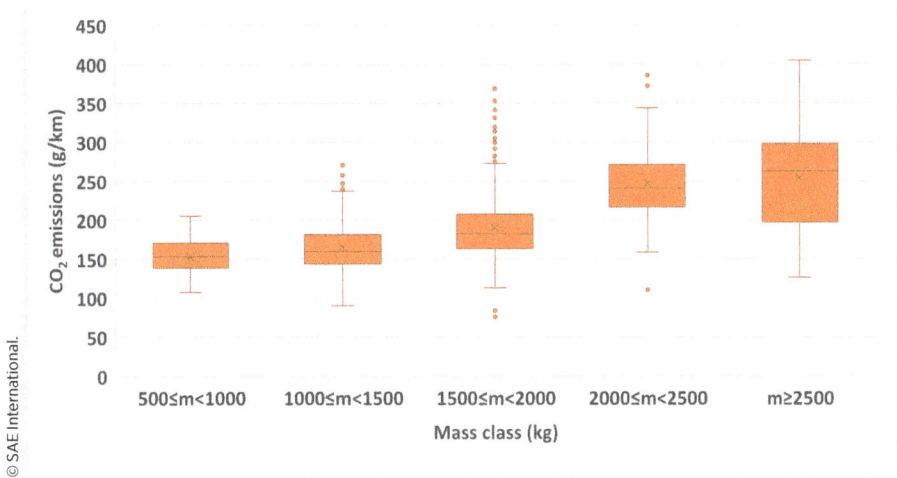

Technicalities understood, what Figure 7.1 shows is that—on average—real-world carbon dioxide emissions clearly rise with vehicle weight. The variability within each mass class is high, but the trend is nevertheless clear. The average emissions of the 1.0–1.5 tonnes (2200–3300 lb) group is 165 g/km, compared to 247 g/km for the 2.0–2.5 tonnes (4400–5500 lb) class, an increase of 50% in carbon dioxide emissions accompanying a mass increase of approximately 80%. Put another way, only 1% of results in the lighter category were *above* 247 g/km, while only 4% of results in the heavier category were *below* 165 g/km. An example of a vehicle that might have lower emissions in the heavier category would be a diesel vehicle. Diesels have emissions approximately 15% lower than the equivalent gasoline vehicle, even though like-for-like diesels are typically of greater mass than their gasoline counterparts because of the heavier engine required to withstand the compression ignition inside the diesel engine. Similarly, a high-emitting car in the light category is often gasoline-fueled. Overall, despite the variabilities, it is clear that this data supports the concept that mass is a good indicator of carbon dioxide emissions.

Regulated Pollutant Emissions

One of the gases that originally inspired air quality regulation—because of its malign effects in suffocating humans—was carbon monoxide [7.21]. Present in ambient air at approximately 100 parts per billion (ppb), at concentrations above 200 parts per million (ppm) (approximately 2000 times higher), carbon monoxide is rapidly fatal [7.22]. From vehicles, it is primarily produced by the combustion of gasoline with a lack of oxygen-bearing air [7.23]. Diesel engines do not suffer the same problem as their combustion occurs with excess air, by design [7.23]. Figure 7.2 shows how these emissions correlate with vehicle mass.

FIGURE 7.2   Carbon monoxide emissions for different ICEV mass classes for 1190 vehicles.

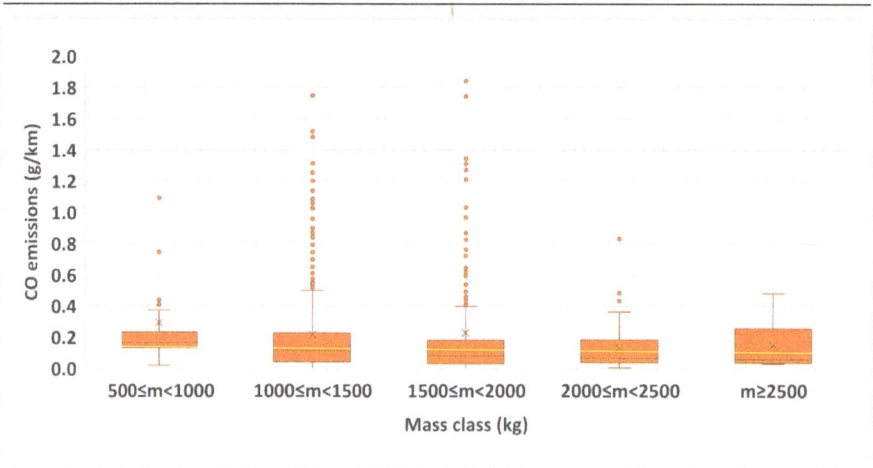

© SAE International.

There is no correlation. If anything, there is a slight tendency for heavier vehicles to emit less carbon monoxide. Not only is there a little trend, but there is also significant variability within each class, especially at the midrange weights. Why might this be? There are two main reasons. First, diesel engines produce relatively little carbon monoxide because of the nature of their combustion described above, and as these engines are higher in torque—they can apply higher forces which are good for heavier applications—they tend to be overrepresented in the higher mass categories.

Second, carbon monoxide is almost entirely cleaned up in the exhaust using aftertreatment called "three-way catalysts," but more colloquially known as "catalytic converters" [7.2]. Good aftertreatment can eliminate almost all the carbon monoxide, but they need to be of a size, quality, and setup (or "calibration") to work effectively. Expensive cars, which tend to be heavier, can accommodate physically and financially the best converters, leading to lower emissions. As a result, we can conclude that carbon monoxide emissions do not correlate with vehicle mass.

Nitrogen oxides are the gases of Dieselgate fame. Although there are multiple gases made up of nitrogen and oxygen, "nitrogen oxides" usually means the sum of nitric oxide (NO) and nitrogen dioxide (NO_2), but excluding nitrous oxide (N_2O)—an anesthetic, extremely potent greenhouse gas, and sometimes a recreational drug [7.24]. Both NO and NO_2 are bad; the latter causes respiratory illnesses in humans, and NO becomes bad as it readily reacts with oxygen to create NO_2 [7.25]. As with carbon monoxide, nitrogen oxides are more powerful as air pollutants than they are as greenhouse gases. Indeed, nitrogen oxides have a global *cooling* potential 150 times greater than carbon dioxide warms the planet! Intuitively, one would expect larger cars, which burn more fuel, to emit more nitrogen oxides, but this is not the case, as shown in Figure 7.3.

FIGURE 7.3   Nitrogen oxide emissions for different ICEV mass classes for 1190 vehicles.

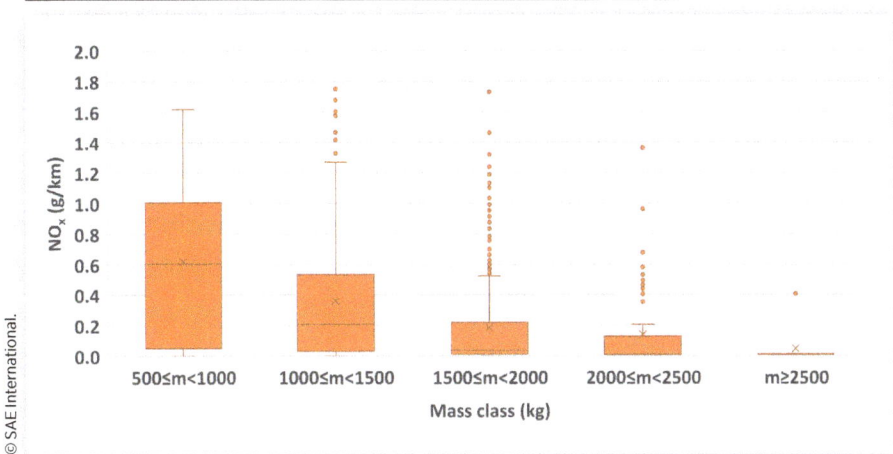

Here, the correlation is strongly negative: larger cars have lower nitrogen oxide emissions. This is not an immediately intuitive result, especially as diesel engines are more prevalent among heavier vehicles, and nitrogen oxide emissions are a symptom of diesel combustion where there is a excess of air relative to fuel in the combustion chamber. Gasoline engines, in contrast, usually have a near-perfect combination of fuel and air, a form of combustion described by the term "stoichiometric." Nitrogen oxides are still produced, but typically at lower levels, and is much easier to clean up because they run stoichiometric. So, why is it lighter vehicles have higher emissions? As with carbon monoxide, nitrogen oxides can be cleaned up very well in the tailpipe, but the technology to do so is even more expensive, and bulkier, than the catalytic converter [7.23]. Prior to Dieselgate, many manufacturers opted for the cheaper "lean NO_x trap" technology for diesel engines, while primarily premium car makers—as well as Peugeot Société Anonyme (PSA), now part of Stellantis—went for the more complex "selective catalytic reduction" (SCR) technology, which injects urea into the hot exhaust to neutralize the NO_x [7.23] from diesel engines. Further, after 2018 and the introduction of the new test regime, all manufacturers opted for the effective SCR system [7.26]. Therefore, the heavier vehicles analyzed above, which are roughly representative of cars on the road today, are dominated by diesel vehicles with SCR and low-NO_x gasoline vehicles, whereas the average emissions of the lighter cars are dragged up by many small, cheap diesel cars with poor aftertreatment systems. So, in a more pronounced way than carbon monoxide, this analysis does not back up the theory that vehicle mass is a good proxy for these emissions.

The final regulated pollutant we will consider is particle number (PN). As the name suggests, this pollutant is not quantified by mass, but by literally a count of the number of particles. This is necessary as particles from a modern exhaust are tiny—in the nanometer diameter range [7.27]. If we were just to use mass as the regulatory metric, no realistic number of particles would add up to a significant mass. However, the billions of nanoparticles coming out of an exhaust can be inhaled and cause significant respiratory and wider health harm in humans [7.27]. Emissions Analytics' dataset for PN is smaller at only 184 tests, and the results are shown in Figure 7.4.

FIGURE 7.4 PN emissions for different ICEV mass classes for 184 vehicles.

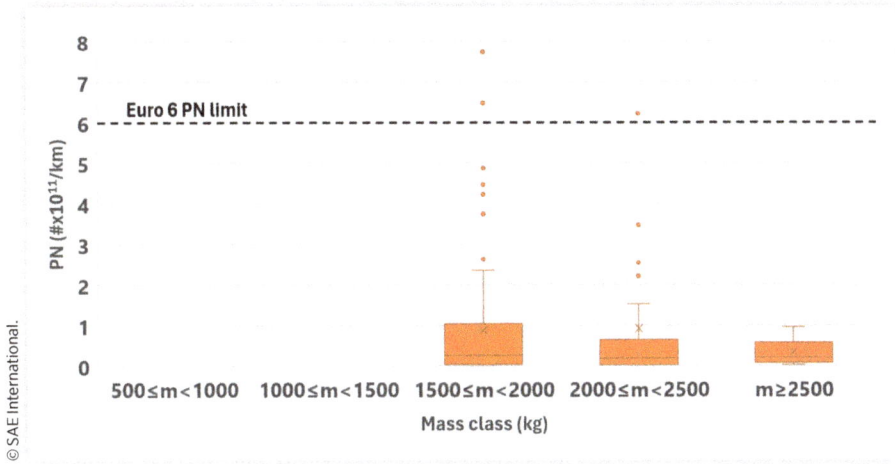

Despite not having any tests of the lightest vehicles and small overall numbers compared to the rest of the pollutants that we have looked at, the trend is again fairly clear. There is a weak downward trend in PN as the vehicle gets heavier, and the variability is much greater in the lighter category. The reasons are similar to the other air pollutants above. Combustion engines create lots of particles, as a result of fuel impurities and incomplete combustion in the chamber, which results in small pieces of primarily carbon being emitted. These are controlled, in modern vehicles, by adding a filter toward the end of the exhaust pipe. Filters—known as either gasoline or diesel particulate filters (GPFs and DPFs) depending on the fuel—are sophisticated honeycombs that trap particles, which are then occasionally cleaned out through a process called "regeneration" [7.28]. Filters come in different shapes and sizes, with different costs. The more expensive the car, which is likely to be heavier on average, the more likely a superior filter is to be installed. In addition, while it has effectively been mandatory to install a filter on a diesel vehicle since about 2009 under the "Euro 5" regulation, this is not the case with gasoline vehicles [7.29]. Only around a half of gasoline vehicles produced in Europe contain a filter now, and the effectiveness—or "filtration efficiency"—of GPFs remains worse than for diesel filters [7.30]. Therefore, the lower PN in the heavier mass classes is due to a combination of better technology on more expensive vehicles, and a higher proportion of diesel engines (we will look at this in detail shortly). Whatever the reason, this again does not support mass as a good proxy for emissions.

All these emissions described are regulated in Europe, the US, and most other locations. The primary exception to this is that the US regulates particle mass but not PN [7.31]. It should also be noted that limits are not set anywhere for maximum carbon dioxide emissions, although there are fleet-average carbon dioxide regulations. Official tests lead to certification values for carbon dioxide that then feed into labeling, taxation, and fleet average emissions limits.

Unregulated Pollutant Emissions

Just because a pollutant is not regulated today does not mean there are not others of concern. So, to complete this section, let us look at a range of currently unregulated gases of potential concern from a health point of view. Figure 7.5 sets out the same averages and variabilities for nitrous oxide (N_2O), formaldehyde (CH_2O), total alkanes, total aromatics, total polycyclic aromatic hydrocarbons, and total alcohols.

FIGURE 7.5   Non-regulated emissions for different ICEV mass classes for 30 vehicles.

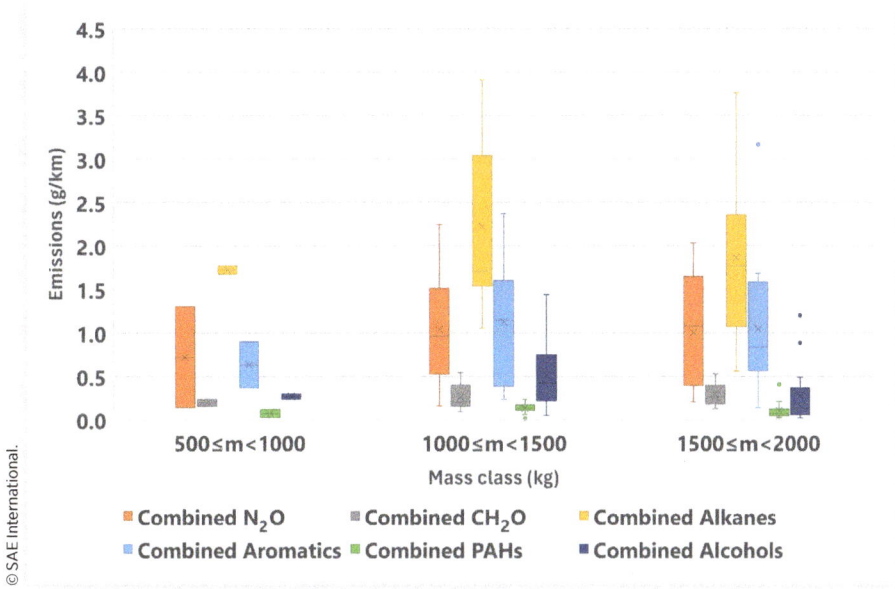

Nitrous oxide is primarily a potent greenhouse gas, approximately 273 times more potent than carbon dioxide over a 100-year period [7.2]. It is also a potential air pollutant in the upper atmosphere, as it may cause stratospheric ozone depletion [7.32]. Over the 30 tests represented in the chart, there is only a weak positive correlation with vehicle mass. Formaldehyde is a known carcinogen, damages human organs, and is a serious irritant [7.33]. Again, there is a weak positive correlation. Alkanes, along with alkenes and alkynes, are the backbone of hydrocarbon fuels and, if inhaled or ingested by humans, can lead to various organ damage [7.2]. There is no correlation between their emissions and vehicle mass. Aromatics, while sounding pleasant, are a group of organic compounds many of which are carcinogenic to humans [7.3]. Again, there is a weak positive correlation. A subset of aromatics are the polycyclic aromatic hydrocarbons, which generally have higher molecular weights and tend to be more carcinogenic [7.3]. Here, there is a weak negative correlation with weight, if anything.

Overall, there is no obvious link between higher vehicle mass and higher unregulated emissions, as we can see in Figure 7.5.

There are, of course, other emissions we could also talk about, but they are of less significance to the overall picture. We could, for example, split the nitrogen oxides down into their nitric oxide and nitrogen dioxide components and look at the mass dependency. We could look at "total hydrocarbons"—a regulatory measure used to quantify the amount of unburnt fuel that makes its way from the tank to the tailpipe because of poor design. Wind back some decades to an era of poorer quality fuel and we could have looked at sulfur dioxide. That problem, however, had to be solved to enable the deployment of catalytic converters without poisoning them [7.2]. We have dwelt on PN above, but there is, of course, also particle mass. US regulators still lead with this metric, as there is most certain evidence as to the health damage from particle mass. In contrast, the medical evidence is that PN may be associated with worse health impacts, although, as a newer research area, that is scientifically less certain. Yet, any vehicle with a particle filter will be registering almost zero for particle mass, as the filters are almost perfectly effective against the larger, heavier particles [7.27].

In short, therefore, in a world of ICEs, vehicle mass is a good predictor of carbon dioxide emissions, but a bad one for almost all pollutant emissions. The deployment of aftertreatment on the vehicle has decoupled the rate of fuel burn from the level of pollutant emissions. If we now introduce BEVs into the analysis, with their zero tailpipe emissions, even the positive correlation between mass and carbon dioxide breaks down as by definition (even if not by reality) they emit zero carbon dioxide. Due to the weight of their batteries, 61% of BEVs on the market now fall into mass categories from 2 tonnes (4400 lb) upward [7.34]. While ICEVs in this weight range have the highest carbon dioxide emissions, their electric counterparts have zero emissions at the (nonexistent) tailpipe.

Looking at the problem of predicting emissions from the perspective of the ICE age, it appears that vehicle mass is a metric of only limited use. As a result, the current patchwork of labeling and taxation as described in Chapter 4 may have some logic. However, the nature of the car fleet is fundamentally changing with electrification. If government policies in developed countries are successful, it is likely that cars on the road will be largely electric by 2050 [7.35]. In that world, tailpipe emissions become entirely irrelevant. We will have a world where emissions are dominated by those from manufacturing and non-exhaust sources. We will move on to analyzing these, and their relationship to mass in the following chapters.

What About Metrics Other Than Mass?

But before that, for completeness, we must consider whether there is another metric that better correlates with tailpipe emissions, including BEVs. Is it possible that there is a superior alternative to mass, despite its centrality in the physical formulae for work done? There are, of course, a wide range of possibilities, but the most obvious contenders would be fuel type, engine size, and Euro regulation stage. Do not forget that only a single metric is allowed. It would be much easier to predict if, for example, you were allowed to know the fuel type, engine size, and Euro stage (the European emissions legislation standard the vehicle was certified under), but that then becomes too complicated for real-world consumer decision-making. We are, therefore, only allowed one.

To recap, the Euro stages are the progressively tighter regulatory limits that have been applied in Europe since 1992 [7.2]. They only apply in legal terms to the air pollutant emissions, not carbon dioxide. Grouping the real-world test into these Euro categories, we can see the relationships between those stages and emissions in Figures 7.6 and 7.7. It is clear that the Euro stage is not a predictor of carbon dioxide emissions, but it is a reasonably good predictor of nitrogen oxides.

FIGURE 7.6 Carbon dioxide emissions for different ICEV Euro standard classes for 1189 vehicles.

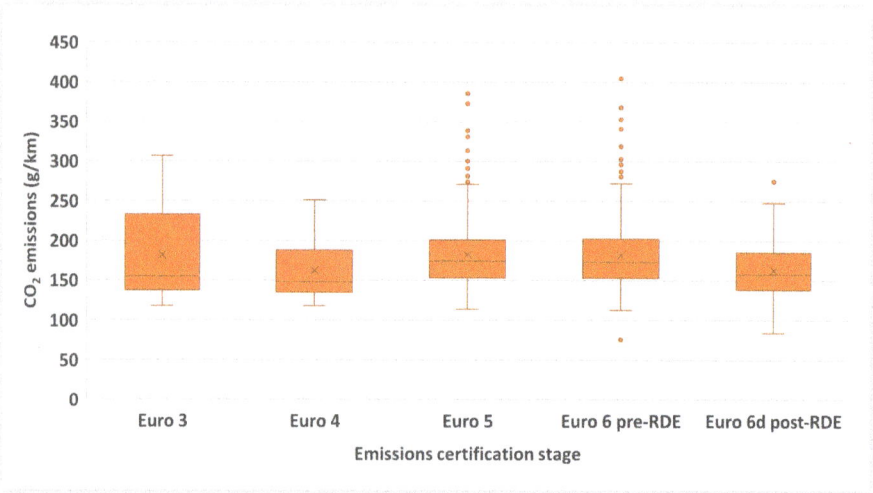

FIGURE 7.7 Nitrogen oxide emissions for different ICEV Euro standard classes for 936 vehicles.

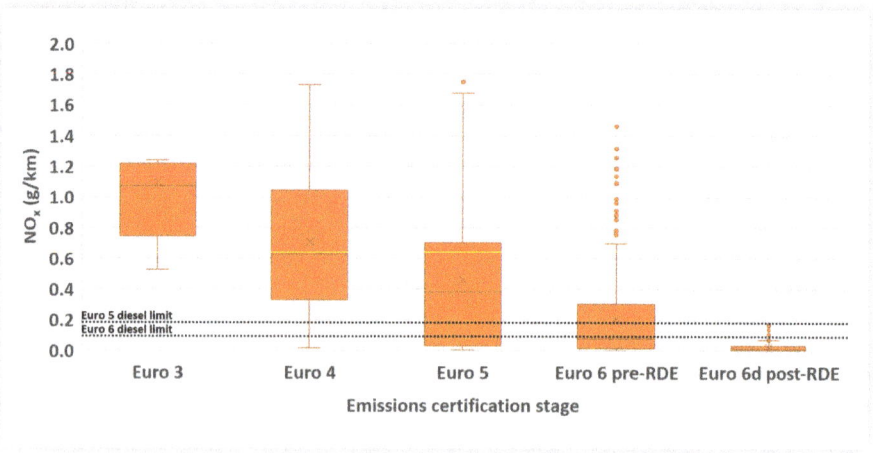

What is remarkable about this last statement is that the Euro stage is not a better predictor of nitrogen oxide emissions. After all, the Euro stages are there precisely to regulate these emissions, among others. From the chart, significant variability can be seen within each Euro stage, which is due primarily to two factors. The first (and innocent) factor is that the vehicles are a mix of gasoline, diesel, and hybrid engines and, as described previously, nitrogen oxide emissions on gasoline vehicles are typically much lower than for diesels and are regulated to different levels. However, this does not explain all the variation. Most of the remaining variability is due to the alleged cheating of Dieselgate. Bear in mind that the maximum emission permissible for Euro 5 was 0.180 g/km and for Euro 6 it remains 0.080 g/km. A high proportion—50%—of diesel cars of the pre-RDE Euro 6 era emitted over the regulatory limit in real-world driving, by factors typically between three and ten times. This has also been seen in other literature [7.36]. Therefore, while the Euro stages give a good guide to these emissions, it can only be muddy and approximate because of this manipulation of the system.

Turning to fuel type, we consider carbon dioxide emissions in Figure 7.8. We are using "fuel type" in quite a loose sense, which may be better described as "powertrain" to take in both the different liquid fuels of gasoline and diesel, and various grades of electrification using hybrid propulsion systems. With carbon dioxide emissions, there is some trend, with gasoline vehicles having higher emissions on average than diesel vehicles—the original reason for the European push for diesel cars in the 1990s [7.2]. Hybridizing gasoline reduces the average emissions by 29%. However, within each of these classes, there is such variability that knowing the powertrain only gives a rough guide to the carbon dioxide emissions.

FIGURE 7.8 Carbon dioxide emissions for different powertrain types for 1189 vehicles.

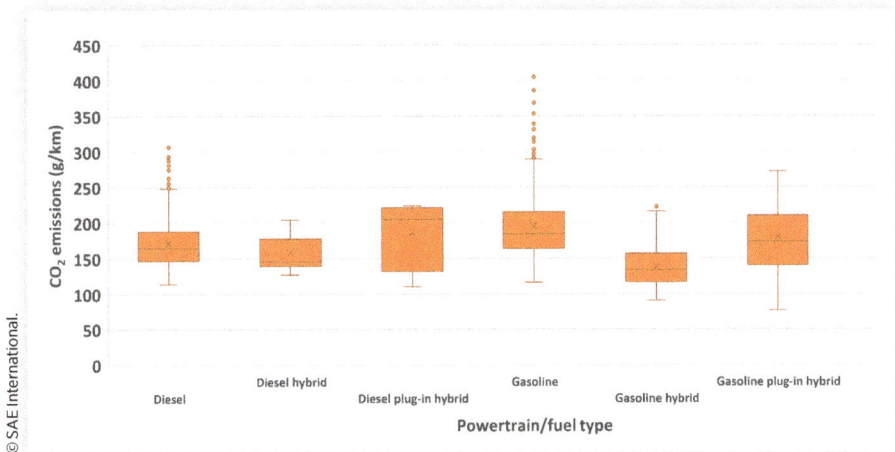

A similar conclusion is drawn with respect to carbon monoxide emissions (Figure 7.9). Realistically, all these segmentations tell you is that gasoline vehicles tend to have higher carbon monoxide emissions than diesel vehicles—which is not big news. The reverse is true for nitrogen oxides—diesel is worse than gasoline (Figure 7.10). For a consumer looking to buy the right car, this data just does not give you a clear steer—whether to a specific powertrain type or vehicle model. PN (Figure 7.11) gives a less clear pattern, but broadly aligns with what we know already—that DPFs work really well [7.26], reducing these emissions to below those of gasoline vehicles on average (most of the gasoline vehicles in this sample are not fitted with a GPF).

FIGURE 7.9 Carbon monoxide emissions for different powertrain types for 1183 vehicles.

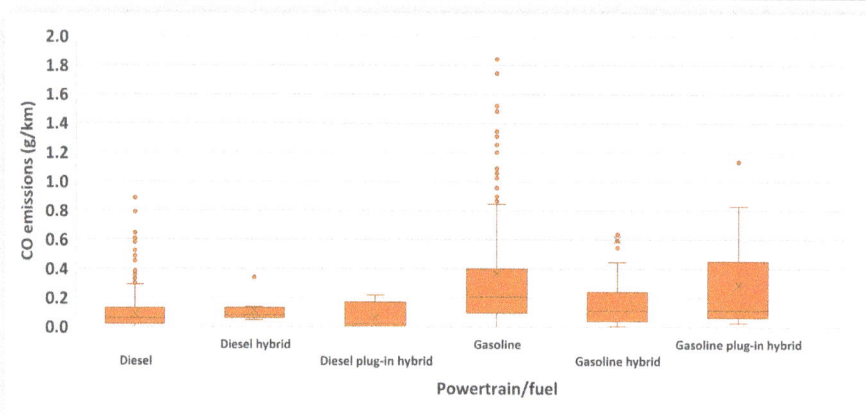

© SAE International.

FIGURE 7.10 Nitrogen oxide emissions for different powertrain types for 936 vehicles.

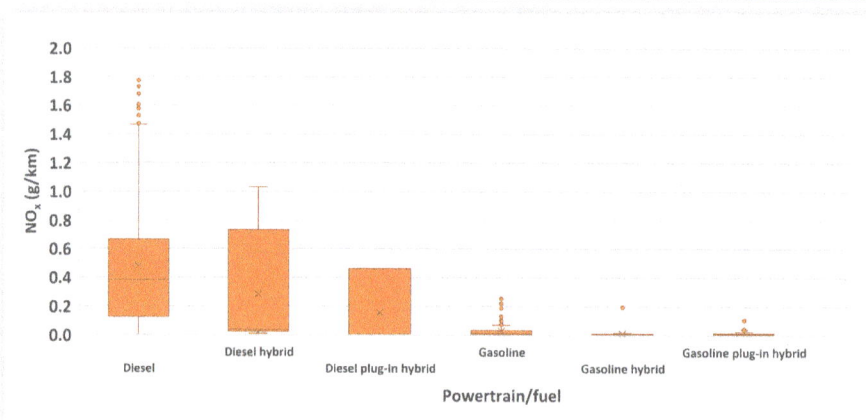

© SAE International.

FIGURE 7.11 PN emissions for different powertrain types for 183 vehicles.

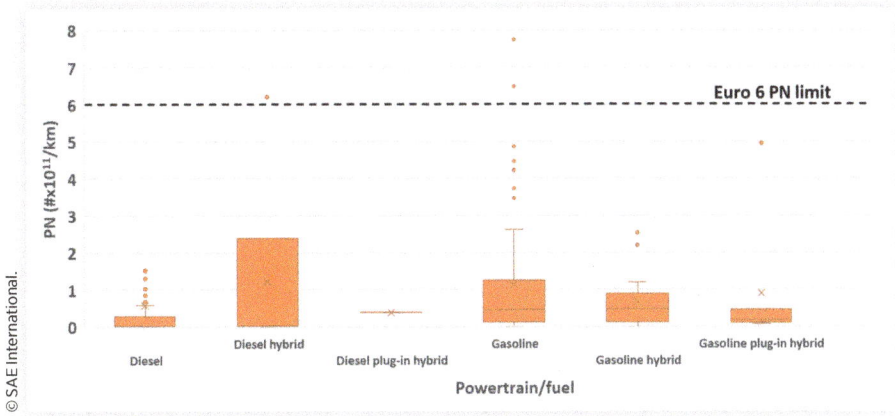

FIGURE 7.12 Carbon dioxide emissions for different engine sizes for 1189 vehicles.

This leaves engine size as the last obvious potential metric to forecast the emissions of a vehicle. Here, the results are more promising, with a strong correlation between the engine displacement—the volume in the cylinders—and carbon dioxide emissions (Figure 7.12). Even though there is large variability in the mid-to-large engine sizes, the trend is both unmistakable and intuitive as bigger engines tend to have more power and consume more fuel. In contrast, the relationship between engine size and nitrogen oxide emissions is nonexistent (Figure 7.13), because of the mix of powertrains within each category, the effect of aftertreatment, and problems with the manipulation of homologation testing discussed above.

FIGURE 7.12 Carbon dioxide emissions for different engine sizes for 1189 vehicles.

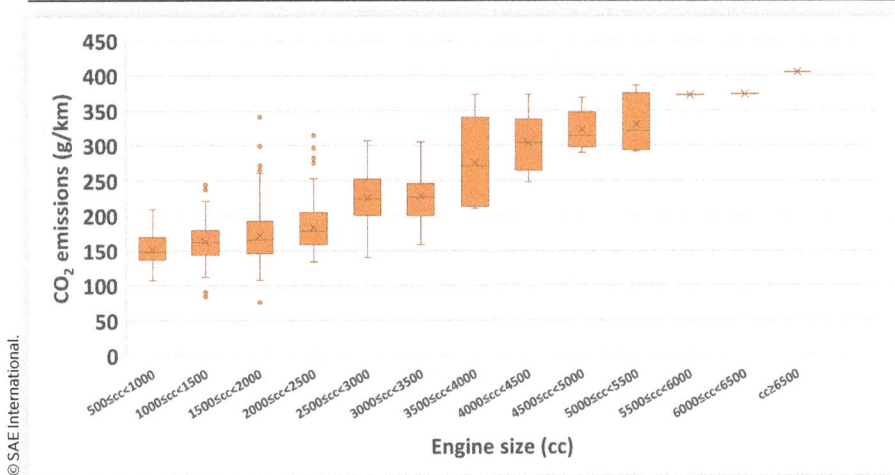

FIGURE 7.13 Nitrogen oxide emissions for different engine sizes for 936 vehicles.

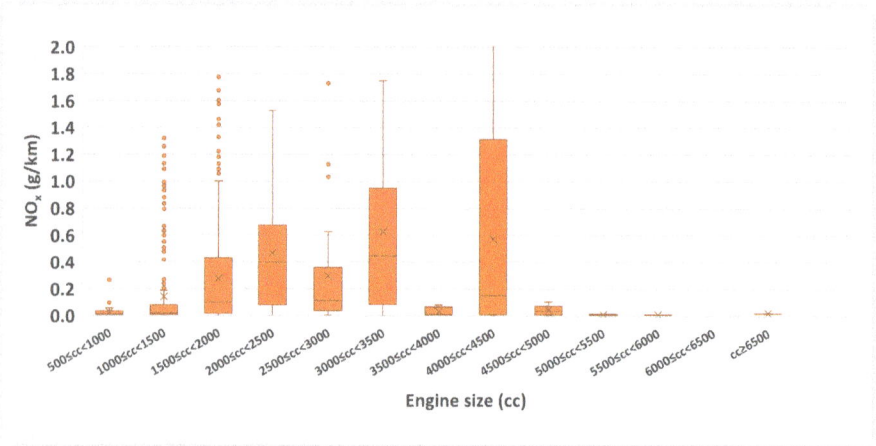

© SAE International.

Summary

We conclude this chapter with the finding that mass is a good predictor for carbon dioxide emissions, but not for other air pollutants, but all the other obvious options are no better, or worse. Engine size is, in fact, a somewhat better predictor of carbon dioxide emissions—with a coefficient of determination (R^2—a measure of statistical association) of 0.41 compared to 0.24 for mass—but is worse for many others. This reflects how the ICE has evolved as a highly complicated piece of technology, with engine optimization reducing the pollutants in the first place and a raft of aftertreatment technologies cleaning up the remaining mess in real time. The interactions between the engine and aftertreatment are complex and nonlinear, with the tailpipe emissions being the net result of a big number being subtracted from a (much) bigger number. Through this process, the levels of tailpipe pollutant emissions are decoupled from the fuel burnt or the work done. The exception is carbon dioxide, where there is no technology for onboard cleanup (because of its relative unreactivity) coupled with the sheer volume of the gas—averaging around 1 kg emitted for every 6 km driven. It is not, therefore, surprising that the correlation remains between vehicle mass and those emissions.

While it would be tempting, because it only holds for carbon dioxide and not any of the other tailpipe emissions, to throw out our hypothesis at this point of our book, we must now consider the life cycle, non-exhaust, and "unconventional" emissions to build a complete picture, in the context of a car fleet rapidly electrifying.

8

Life Cycle Emissions

IN THIS CHAPTER

- Tailpipe emissions are not sufficient to determine the environmental impact of a car.

- Life cycle analysis shows that weight is also a good predictor of both life cycle carbon dioxide emissions and wider environmental impact.

- Options that reduce weight but increase emissions, such as extreme lightweighting materials, are typically of a cost that restricts them to niche applications.

n 1980, Milton Friedman, the famous economist, made a now much quoted statement: "Look at this lead pencil. There's not a single person in the world who could make this pencil" (**Figure 8.1**). He then went on to discuss growing and cutting down the tree to obtain the wood. Cutting and processing that wood to make the outer part of the pencil. Graphite for the "lead" is mined in South America, transported to the point of manufacture, and processed into shape. The brass for holding the eraser in place (brass would start as zinc and copper ores). The eraser itself (derived from crude oil most likely). The paint for the pencil. Then shipping the manufactured pencil to the point of sale [8.1].

FIGURE 8.1 A pencil.

Vitaly Zorkin/Shutterstock.com.

You get the idea, even the simplest products that we can think of have immensely complex supply chains and no one individual could conceivably create them themselves.

Now imagine trying to work out the pollution impact of that pencil. There would be carbon dioxide emitted at almost every phase. What about the use of the land on which the tree was grown? Are there other emissions associated with mining raw materials, such as leaching ores, dust, or other pollutants? The challenge of calculating the environmental impact of a single pencil is equally large, if not larger than manufacturing it itself.

The cars of today are miles ahead of the pencil of 1980 in terms of complexity.

How can we cope with this? The answer is life cycle assessment.

LCA is an approach that breaks down all the phases of a vehicle's existence and considers the environmental impact of each [8.2]. Figure 8.2 illustrates the overall concept and phases considered, including obtaining the raw materials for manufacture, their processing, production of the vehicle, production of the "fuel" (electricity, gasoline, whatever), use of the vehicle, and, finally, its disposal and recycling [8.3]. We have covered the in-use phase elsewhere in this book, so this chapter will focus on all the other aspects.

FIGURE 8.2   The phases considered in LCA [8.3].

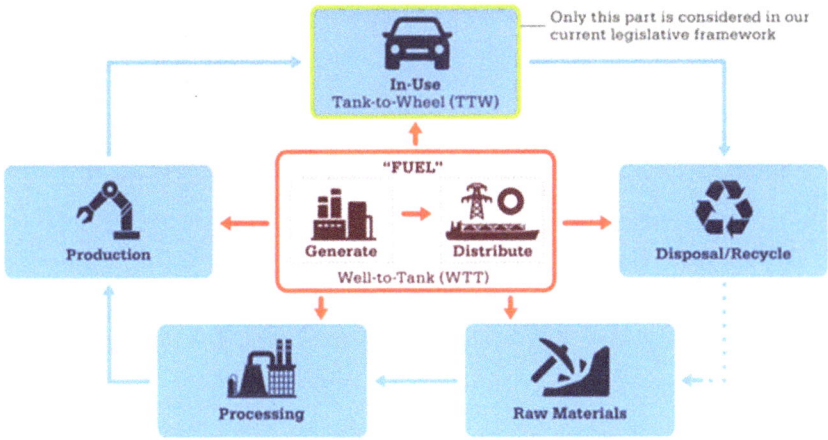

The first question to consider is *why bother*? We have not needed LCA for 150+ years of car use; why should we today?

Well, for those 150 years we have not always cared about the environmental impact of cars. Initially, motor vehicles, containing an ICE, were seen as such an improvement, both in terms of functionality and environmental impact, over horses, which were the dominant mode of transport then. That improved functionality allowed more to be carried over greater distances. It also significantly reduced disease in cities from the reduction in horse poo [8.4].

For around a hundred years—let us say 1910 to 2010—ICE-powered motor vehicles dominated transportation [8.5]. The environmental impact of these vehicles is overwhelmingly from their tailpipes and the in-use phase. Almost all studies, no matter what the use-case or input parameters, agree that over 75% of the life cycle emissions—whether that is carbon dioxide, pollutant emissions, or other things—come from the tailpipe and the use of the vehicle itself [8.2].

Another good reason for considering only the tailpipe of a car is that you have a single measurement point. Anything that passes through that single location is counted and everything else is not. This means that you can put some fancy (and expensive) instrumentation at that single location and measure, with a high level of accuracy, a wide range of pollutants and greenhouse gases. Easy. These reasons are, of course, why almost all emissions regulation has focused on the tailpipe, as we have seen.

As we move to this new world of transportation, so we need to change how we assess the environmental impact of cars. The tailpipe is no longer enough. BEVs do not have a tailpipe—does this therefore mean that they emit nothing, as many people say? No. They may have lower in-use-phase emissions than their ICE counterparts, but their emissions are not zero [8.6]. Their emissions simply come from places other than the tailpipe. This is why we need LCA.

Of course, moving away from the single measurement location to multiple (often hundreds or even thousands) measurement locations adds substantial complexity, uncertainty, cost, and data requirements. Suddenly you need to be able to assess emissions associated with mining raw materials, transporting them, processing them into preproduction material, shipping them again—you get the picture. It is clearly impossible (well, perhaps nearly impossible—certainly impractical) to make all these measurements. It is also challenging to apportion exactly which bit of raw material or electron in a wire ended up in your car. Therefore, assumptions need to be made, and the nature of these assumptions has a huge impact. Before we get there, however, let us look at the first efforts to move away from solely tailpipe emissions.

The first steps in this direction began in the late 1990s and early 2000s as biofuels began to make an impact on the fleet. These first-generation biofuels (ethanol was by far and away the most common biofuel for cars), fuels made from plants such as corn, at least saw carbon dioxide absorbed as the plant grew. This meant that simple tailpipe carbon dioxide analysis that was (and still is) the favored assessment methodology did not capture the benefits of these fuels. So, it became important to take into account the carbon dioxide that was emitted (and, importantly, absorbed) during the fuel production. This led to so-called "well-to-wheel" analysis (WTW) [8.7].

WTW

WTW analysis takes into account all emissions associated with the fuel before it is combusted in the engine and emitted through the tailpipe, and then adds tailpipe emissions to set the total. In a traditional fossil fuel ICEV, this comprises the emissions from the oil well to the wheels of the vehicle, where the power is generated— and hence the vehicle moves. This allows for some negative carbon dioxide emissions, if there are any, associated with cultivating a plant, which absorbs carbon dioxide from the atmosphere as it grows, to provide your biofuel. It also begins to deal in a more comprehensive way with the true carbon dioxide emissions from fueling a fossil fuel-powered car, such as extracting and refining the oil and transporting it to the consumer [8.8].

Of course, WTW analysis does not just have to be applied to biofuel vehicles, and recently, this analysis has been applied to BEVs [8.8], hydrogen vehicles [8.9], and even e-fueled vehicles [8.10]. E-fuels are a synthetic energy source made from captured carbon dioxide using renewable electricity; they have significant potential to decarbonize the existing fleet [8.11]. This WTW approach makes sense in some ways; the environmental impact of using a vehicle is clearly a function of how much it is used—that is, how far it travels. Of course, all vehicles have a certain fuel use per distance travelled and this can be relatively easily normalized over all vehicles of the same powertrain. As an example, it is clearly fair to compare the fuel usage of different ICE cars as they all achieve broadly the same task, and their fuel use will be inextricably linked to their carbon dioxide emissions. Similarly, comparing the electricity use of different electric cars is clearly a sensible thing to do [8.12].

However, when comparing different powertrains, for example ICEV vs BEV, this approach does not accurately take into account the environmental impact of each car. This is because the emissions associated with manufacturing are substantially different between these different powertrains [8.5]. To do this properly, we must use LCA.

Full Life Cycle

As discussed previously, due to the challenges associated with acquiring such a large amount of data about the inputs to an individual car, the assumptions that need to be made in an LCA are manifold. Indeed, this can lead to widely varying and conflicting LCAs [8.5]. Common assumptions that need to be made, among others, are:

- The carbon dioxide intensity of the energy used in manufacturing the vehicle.

- The equivalent intensity of the electricity used to charge the vehicle, if it plugs in. Do you use grid averages or some form of marginal value? A marginal value is the extra emissions emitted when an individual source is added to the grid—such marginal additions are often serviced by gas or coal plants rather than nuclear and renewables, which provide the base load [8.13].

- The carbon dioxide intensity of the raw materials.

- The carbon dioxide intensity of the liquid fuel for ICEVs, including considering whether it uses bio- or e-fuels rather than fossil fuels.

- The carbon dioxide intensity of the land use associated with biofuels and mining raw materials [8.14].

- The vehicle lifetime.

- The vehicle lifetime distance driven.

- The vehicle powertrain.

- The vehicle battery size.

- Vehicle maintenance, including significant parts that might need replacing, for example batteries.

- Climate—for example, if the car is used in a warm or cold country.

- Geography—what is the grid you are charging from, and are there other local factors (for example terrain, access to infrastructure) that are relevant?

Often, future predictions of many of the above need to be made too—for example about future electricity grid decarbonization. This, of course, adds even further uncertainty to any LCA that attempts to do that.

Does this mean that we should throw up our arms and say that LCA is pointless? Quite the reverse. LCA is vital. However, for it to be used properly, it is important to be transparent and up-front about the details. Some things to look out for include a clear definition of the remit (what is in and out of scope for the particular LCA), what values it uses for all components (carbon intensity of electricity, battery manufacture, or whatever—see the long list above), what it compares (for example comparing a 20-year-old ICEV to a hypothetical future BEV will give one result, and comparing a light-weight e-fueled hybrid to a large extended-range BEV pickup charged on coal will likely give the opposite result), and whether or not the data used are verifiable and open source (or do they rely on confidential internal company information). Fundamentally, if enough information is there to enable a user to recreate the particular LCA and put some of their own numbers and assumptions in, then it is probably a good starting point [8.5].

Whole books could, and indeed have, been written on the topic of LCA, but this is not that book [8.15]. There are some standards, albeit quite method-based rather than applied, such as ISO (International Organization for Standardization) 14040 and 14044, which set standards for LCAs. We would note that few LCAs appear to be compliant with these today [8.16].

Many LCAs have been undertaken over the years—a Google Scholar search for "life cycle assessment AND car" gives nearly 60,000 results at the time of writing. A similar search for "LCA ICEV BEV" (using the common acronyms for the propulsion technologies of interest) returned 2680 articles in 2020 and 9510 articles in 2024 [8.5]. This is an active and fast-growing field.

How, then, are these studies undertaken? Mostly, these are "paperwork" exercises where there is a particular variable of interest (from the long list above) that is varied, and averages are taken of the others and held constant. Think of a simple spreadsheet analysis. There is a wide range of literature values that are available for these various parameters, and people who conduct such studies typically analyze the literature and choose, by some metric or another, values that are expedient for their study. This is, of course, where potential bias, or perhaps better expressed as perceived or even unconscious bias, can come in. The selection of these values effectively determines the outcome of the LCA, and this is how seemingly contradictory results can arise. This does not make a particular LCA "right" or "wrong," but it does, however, make the transparency as to the selection of these values all the more important. What is that LCA trying to achieve, where is it based, and do you, the reader, consider the authors' choices for values justified? Are there particular financial or political interests behind the choices?

More recently, in the interests of ease and transparency, online tools have become available to assist people undertaking LCAs. These tools typically "hard code" some of the analysis to enable simpler calculations. They do, however, enable sensitivity analyses of parameters to be undertaken more easily, as well as forcing a user to engage with the choice of values rather than just accepting the defaults. A popular example is the cars carbon dioxide comparator (https://www.carsco2comparator.eu/), which was developed by IFP Energies nouvelles (IFPEN), a French public research organization [8.17]. This allows users to select from a range of powertrain types, vehicle lifetimes, battery carbon dioxide intensities, and other parameters to quickly obtain an LCA. Such tools, which are freely and publicly available, and allow users to make their own decisions, are a powerful addition to the field and help make LCA more accessible.

At the extreme end of the spectrum, a largely hard-coded tool to assist with LCA, and probably the "gold standard" when it comes to automotive sector LCAs today, is a software package called **GREET** (The **G**reenhouse Gases, **R**egulated **E**missions, and **E**nergy Use in **T**ransportation Model), which was developed by the Argonne National Laboratory in the US, funded by the US Department of Energy [8.18]. GREET is complex modeling software that allows many parameters in an LCA to be varied easily, while providing default values that enable practical LCAs to be undertaken by a wide range of users.

For a set of input variables and cars of interest, GREET will output the energy usage, whether from fossil or renewable sources, emissions of three greenhouse gases (carbon dioxide, methane, and nitrous oxide), and so-called "criteria pollutants" (carbon monoxide, nitrogen oxides, sulfur oxides, particulate matter [PM_{10} and $PM_{2.5}$ separately], and volatile organic compounds). While this is not an exhaustive list, these nine emissions are among the most important for any LCA applied to cars.

Such is the status of the GREET model that it is now incorporated in law. It was specified in the US Inflation Reduction Act (2022) as the appropriate tool to calculate life cycle greenhouse gas emissions [8.19]. The results are used to determine the tax credit given for clean hydrogen production. In addition, it has been used by the California Air Resources Board (CARB) to help develop its Low Carbon Fuel Standard (LCFS) [8.20]. Like any tool, however, the use of GREET in and of itself does not guarantee success or indeed quality. In addition, as a much more complex hard-coded tool, its use requires specialist skills, and so, unlike the online tools above, it is designed for experts rather than accessibility.

With this brief introduction to LCAs, how do they apply in our context? If LCA is the most useful, and possibly only, fair way of comparing vehicles with different powertrain technologies, then we must see what the effect of vehicle mass is in an LCA. As the previous chapter looked at the in-use phase, this analysis will consider the other parts of the LCA.

In the discussion above, our focus has been on greenhouse gas emissions as this is clearly the parameter of most interest given climate change concerns. However, there are other environmental impacts that are worth noting and should be taken into account. Typically, the production and disposal phases significantly contribute to human toxicity potential (HTP), mineral depletion potential (MDP), and freshwater eco-toxicity potential (FETP). On the other hand, the in-use phase dominates for Global Warming Potential (GWP), terrestrial eco-toxicity potential (TETP), and fossil depletion potential (FDP).

HTP is a catch-all metric for things that are, or can be, toxic to humans. These include air pollution (for example, carbon monoxide and particulate matter), heavy metals (of these, cadmium, lead, mercury, and nickel are of particular interest for propulsion technology applications), dust, carcinogenic hydrocarbons (many, but not all, hydrocarbons are carcinogenic), and possibly even radiation. HTP is often expressed in units of kg of 1,4-dichlorobenzene (1,4-DB) equivalent—1,4 dichlorobenzene being a toxic and probably carcinogenic chemical commonly used as a deodorant, disinfectant, and pesticide, most familiarly in household mothballs.

MDP, sometimes referred to as abiotic depletion potential (ADP), is the potential to deplete finite sources of scarce raw materials. Similarly, FDP is the potential to deplete fossil sources such as coal and oil. FETP refers to the potential for toxic matter to contaminate fresh waterways, while TETP is the equivalent for contaminating land. Terrestrial acidification potential (TAP) is the potential to acidify the land.

For cars, HTP, MDP, and FETP are caused primarily by the supply chains involved in the production of the vehicles. The in-use phase typically has the highest impact for TETP and FDP. End-of-life adds only a marginal contribution across all impact categories. Studies show that the production of BEVs increases all these categories more than the production of ICEVs except TAP [8.21].

When thinking about applying LCAs to the context of this book, it is worth splitting the car into its constituent parts: the glider (car without propulsion or energy store), energy store (battery, gasoline tank, hydrogen tank, and so on), and powertrain (engine, fuel cell, or electric motor).

Glider

The glider is the main part of the car, consisting of the chassis, body, wheels, and other parts. In short, the whole car without the energy store or powertrain. This part is common to all vehicles regardless of their propulsion system. When it comes to LCA, this part should be relatively consistent between vehicles in terms of its environmental impact. There are only a few things that can be changed within the glider to affect its mass or its impact.

The first, and most obvious, is that the size of the glider will clearly have a big impact on both its mass and its environmental impact. A smaller glider will have lower mass and environmental impact from its production. That is simply because there will be fewer constituent components making it up—less steel and aluminum, fewer plastics, less everything [8.22]. Given fewer input materials, there will clearly be a lower environmental impact from their production. As an aside, there will also be reduced in-use phase emissions as less mass is being transported around, as we have seen many times in this book.

That is an important, if obvious, result. Cars do need to be of a certain size to be useful. Moreover, consumer preferences show a desire for larger cars, indicating that they have more utility, feel safer, and have higher status. There is a long list of small cars, both electric and ICE powered, which have had a small (sometimes cult) following but have ultimately not made much headway in the market beyond the small city-car segment—see Figure 8.3, how many do you recognize? The most successful car of this type has probably been the Smart Fortwo, which sold 1.7 million units worldwide in the period 1998–2015 [8.23]. Contrast the Smart Fortwo, which had almost no competition in its segment, with the Ford Focus, one of many popular cars in the compact segment, which sold in excess of 12 million units worldwide in the same period [8.24].

FIGURE 8.3 Small cars that have not made a big market impact (clockwise from top left, Mitsubishi I-MiEV, Smart Fortwo, REVAi / G-Wiz, Tata Nano).

Sergey Ryzhov/Shutterstock.com.

Grzegorz Czapski/Shutterstock.com.

(Continued)

FIGURE 8.3 **(Continued)** Small cars that have not made a big market impact (clockwise from top left, Mitsubishi I-MiEV, Smart Fortwo, REVAi / G-Wiz, Tata Nano).

Andrei Kuzmik/Shutterstock.com.

Art Konovalov/Shutterstock.com.

A corollary of the link between mass and environmental impact is that there will be a benefit in creating more incentives for people to buy and drive smaller vehicles. However, we must acknowledge that simply shrinking cars is not the only solution. It is also not practical in some cases: to do certain things, a vehicle must need to be of a certain size.

However, there are other things that can be done to make a vehicle lighter. Lightweighting, the act of making an existing vehicle functionally the same but lighter weight, has been a feature of the automotive industry since the 1970s, when the oil price shocks happened, but really accelerated rapidly in the 2000s [8.25]. This is achieved by using advanced materials that attempt to combine lower density (and hence mass per unit size) and higher strength (so you need to use less of them to do the same job).

A good example of this is replacing steel components in a vehicle with aluminum ones—which has a density roughly 2.5 times lower than steel. Although aluminum is weaker than steel, if the component is not particularly structurally important—such as a body panel—then the weight savings can be substantial. In a typical car, replacement of key steel components with aluminum leads to a weight saving of 20–30% [8.26].

A more extreme example is carbon-fiber-reinforced plastic (CFRP)—often known simply as carbon fiber. This is commonly used in high-performance road cars and race vehicles such as Formula 1 (F1) cars. CFRP replacement of steel parts can commonly lead to weight savings of approximately 50% [8.27]. Of course, this comes at a price, with CFRP being significantly more expensive than steel and aluminum, and hence its use is restricted to these high-performance vehicles.

Decisions on materials for a vehicle are driven both by utility (for example, use of CFRP in F1 cars, for which being lightweight is the dominant consideration) and cost [8.28]. The cost of materials varies quite widely, but as an idea, steel typically trades in the £600–£1200 ($750–$1500) per tonne range (at the time of writing), depending on exactly the type of steel you want. Aluminum prices have been volatile recently, but are in the range £1600–£2800 ($2100–$3500) per tonne. CFRP is very expensive, with prices in 2023 being £24,000–£67,000 ($30,000–$85,000) per tonne depending on the grade required. Of course, one of the reasons for using these less dense materials is, well, that you require less of them by mass, but nevertheless, these price differences are stark and it is easy to see why many gliders are predominantly constructed out of steel.

These three materials each have their own environmental foot-print. For 1 tonne of steel, approximately 1.4 to 2.75 tonnes of carbon dioxide (equivalent) is emitted when manufactured [8.29, 8.30], while making 1 kg of aluminum can emit in the range of 7 to 20 tonnes of carbon dioxide (equivalent) per tonne of aluminum [8.31]. Carbon dioxide (equivalent) is a unit of measurement that takes into account the total greenhouse gases (which may include things other than carbon dioxide, such as methane or nitrous oxide) emitted by an activity or process and accounts for them all as the equivalent of the same amount of only carbon dioxide being emitted to allow for easy comparisons. CFRP, being made of carbon itself, has the highest environmental footprint, emitting in the range of 20 to 25 tonnes of carbon dioxide (equivalent) per tonne of CFRP [8.32].

These environmental footprints clearly have large ranges on them, particularly for aluminum, which is due to different approaches to making them as well as the geographic location where they are made. The processes are all electricity-intensive, and different electricity grids worldwide have different carbon intensities, because of the fuels and processes used. In the case of steel, whether it is produced in a blast furnace (where a form of coal is used to heat iron ore) or an electric arc furnace makes a big difference. Aluminum production is even more energy-inten-sive, involving generating alumina from bauxite (the ore) and then electrolyzing it through the Hall–Héroult process—hence its very high environmental footprint. These processes can have their carbon intensity reduced by using renewable electricity, but today, they are largely not. Given the reliance on coal and electricity, you can see that the geographic origin of the material will have a big influence on its carbon dioxide emissions too, especially with respect to the carbon dioxide intensity of the local electricity grid.

For non-carbon dioxide pollutants, research into the effects of steel, aluminum, and CFRP production on HTP, MDP, FETP, and TETP is more limited. What evidence there is in the literature indicates that for aluminum these parameters are mostly dominated by the power (electricity) production [8.33]. For steel it strongly depends on whether a blast or arc furnace is used and FETP seems to be significant [8.34]. CFRP has very little information available but what limited information that there is indicates that its impact may be significant [8.35]. In this case, FDP will track the carbon dioxide emissions relatively closely.

Energy Store

The energy store on a car is the first area where we will see significant differences between powertrains. For a liquid-fueled ICEV, the embodied carbon dioxide in the energy store is almost negligible regardless of its mass—the embodied carbon dioxide in the plastic used to make the tank over the lifetime of the vehicle will be almost zero relative to the mass of the whole vehicle. This contrasts, of course, with the gasoline or diesel itself, which has high embodied emissions. The weight of the fuel (approximately 50 kg when the tank is full), which is included in the kerb weight figures we are using in this book, will play a minor role in the total weight of the vehicle (approximately 3% for a full tank) and hence a minor role in the associated emissions *from carrying that extra weight around* (as distinct from the emissions of burning that fuel).

At the other extreme, storage of hydrogen in a vehicle is quite complex. Today's hydrogen vehicles can convert the stored gas onboard either through a fuel cell (most common) or an ICE. Fuel cells work like electrolysis in reverse: hydrogen is fed to an anode and air to a cathode, which are separated by an electrolyte and often a membrane with a catalyst present. The hydrogen ions and their electrons take different paths through the fuel cell (ions through the electrolyte and electrons through a wire) creating electricity, which is then usually buffered in a battery [8.5, 8.36]. Hydrogen ICEs are under development as a way of using hydrogen at higher power densities and using existing infrastructure [8.37, 8.38]. Most hydrogen vehicles today opt for a 700-bar pressurized gaseous storage system. Taking the Toyota Mirai as an example (see Figure 8.4), its tanks hold 5 kg (11 lb) of hydrogen and have an empty weight of 87.5 kg (193 lb) [8.39]. To have the strength, crash-resistance, and smallest possible weight, they are made of CFRP, which as we have seen above has a high carbon dioxide footprint.

FIGURE 8.4   Toyota Mirai hydrogen tank (in yellow) and fuel cell.

Karolis Kavolelis/Shutterstock.com.

In BEVs, the energy store is the battery pack, which makes up a substantial part of the vehicle both in terms of size and mass [8.40]. Usually, the battery pack is rather like a tray that sits under almost the entire vehicle (see Figure 8.5). The battery pack is heavy, so keeping that mass low is important for vehicle handling.

FIGURE 8.5   BEV showing the battery pack taking up almost the entire floor of the vehicle.

Herr Loeffler/Shutterstock.com.

As there is relatively little more that can be done to improve the energy efficiency of a BEV, the only way to boost vehicle range is by putting in larger batteries [8.40]. BEV battery packs are usually made up of many small cells (see Figure 8.6); for example, the Tesla 100 kWh battery pack that powers the Model S and X vehicles consists of 8256 total cells of the type shown in Figure 8.5, arranged in 16 modules [8.41].

FIGURE 8.6   Individual Tesla battery cell used in their BEV batteries.

Grigyovan/Shutterstock.com.

Given that one of the main consumer hang-ups, rightly or wrongly, about BEVs is range anxiety, manufacturers are keen to make their vehicles as long range as possible. This is achieved by putting sizeable (and heavy) battery packs into their vehicles. The storage capacity (in kWh) and weight (in kg) of battery packs in a number of popular BEVs today are shown in Figure 8.7 [8.42]. Of particular interest is that not only does the weight of packs vary hugely (from 1200 kg up to ~3000 kg [2650–6610 lb]), but also that the mass per kWh—which one might assume should be relatively constant—varies widely, from 22 kg/kWh up to 75 kg/kWh. However, it can be noted that the highest kg/kWh values are for the smallest batteries, which might be expected as there are some economies of scale in the battery packaging.

FIGURE 8.7 Battery pack weights for a number of BEVs available today [8.42].

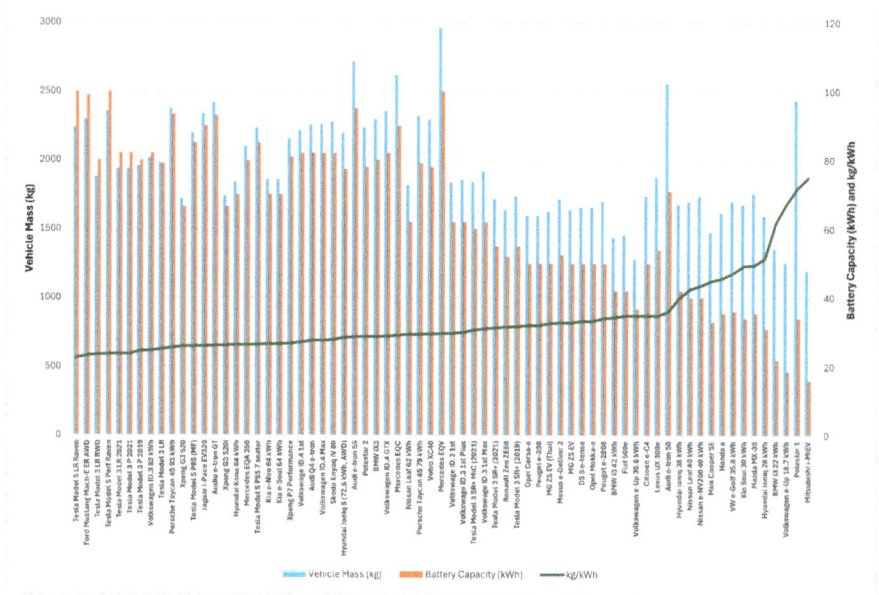

Manufacturing a battery pack has a substantial environmental impact. The raw materials such as lithium, cobalt, and nickel need to be mined, and this requires enormous amounts of energy as well as chemicals and water. Synthesizing these ores into metals, just like with steel and aluminum previously, requires a sizeable amount of energy too—often done at temperatures of approximately 1000°C. When you put all of this together, values for battery pack carbon intensity vary widely, for many of the reasons already discussed in this chapter, but typically lie somewhere in the range of 30 to 200 kg CO_2 per kWh [8.43]. Again, as with the metals we looked at previously, this will strongly depend on where the pack was manufactured, and the electricity grid used to do so [8.44]. This is why decarbonization of grids is such a priority. When scaled to a vehicle level, for a typical, say, 80 kWh BEV battery, these ranges would imply emissions somewhere between 2400 and 16,000 kg CO_2 per vehicle.

As a result of the cost and scarcity of the materials above, there is a trend away from lithium-ion batteries and toward alternative chemistries such as lithium iron phosphate (LFP). These avoid the need for cobalt and nickel, and rely on the more plentiful iron and phosphate. This is likely to reduce the emissions associated with manufacturing a battery – especially the extraction of and refining the material inputs. However, the batteries produced are typically less energy dense, which means that either driving range is reduced for the same weight of vehicle, or the vehicle gets heavier to accommodate a bigger battery to deliver the same range. The former is a compromise on utility, while the latter will lead to increased emissions of other types, including from electricity production, due to the extra weight.

For non-carbon dioxide pollutants, the effect of a liquid fuel tank on HTP, MDP, FDP, FETP, and TETP will be negligible. When considering hydrogen, the high use of CFRP in storage means that the negative effects we observed in the previous section will apply here as well [8.35]. When it comes to battery production, the impact on HTP, MDP, FDP, FETP, and TETP is substantial but with a high variability. This impact depends on how and where the battery is built. In many cases, the impact of battery production alone on these parameters is far higher than the total life cycle impact of other powertrain technologies, but it can be reduced with effective battery recycling [8.44]. Table 8.1 shows the impact of battery manufacture on many of these parameters [8.44]. The important thing to note here is that the functional unit is kWh, meaning that the bigger (and therefore heavier) the battery, the greater the impact on HTP, MDP, FDP, FETP, and TETP is going to be.

TABLE 8.1 LCA impacts of battery production [8.44].

		Per kWh	Battery production (quartile)		
			1st	2nd	3rd
Energy	Primary energy demand	kWh	1.98×10^2	2.84×10^2	5.19×10^2
Global warming	GWP	kg CO_2-eq	7.06×10^1	1.20×10^2	1.75×10^2
Resource depletion	ADP	kg Sb-eq	4.86×10^{-2}	5.93×10^{-2}	4.22×10^{-1}
	Water depletion	m^3	2.37×10^{-1}	5.62×10^{-1}	5.70×10^{-1}
	FDP	kg oil-eq	4.02×10^1	5.08×10^1	6.89×10^1
	MDP	kg Fe-eq	9.10×10^1	9.99×10^1	1.51×10^2
Acidification	AP	kg H+ Mol-eq	1.50	2.33	4.41
	TAP	kg SO_2-eq	1.77	2.03	2.65
Ecotoxicity	FWAE	kg 1,4-DB-eq	8.48	9.61	9.91
	MAE	kg 1,4-DB-eq	9.42	1.04×10^1	1.07×10^1
	TE	kg 1,4-DB-eq	4.45×10^{-2}	4.95×10^{-2}	5.15×10^{-2}
Eutrophication	EP	kg N-eq	9.26×10^{-3}	8.21×10^{-2}	2.08×10^{-1}
	MEP	kg N-eq	1.37×10^{-1}	2.41×10^{-1}	3.00×10^{-1}
	TETP	mole N-eq	1.23	2.12	4.96
	FEP	kg P-eq	1.60×10^{-1}	1.83×10^{-1}	2.95×10^{-1}
Human toxicity	HTP	kg 1,4-DB-eq	1.53×10^2	2.50×10^2	3.82×10^2
Ozone	ODP	kg CFC 11-eq	1.10×10^{-5}	1.16×10^{-4}	1.07×10^{-3}
Air pollutants	PMFP (2.5)	kg PM2.5-eq	1.39×10^{-1}	1.62×10^{-1}	1.99×10^{-1}
	PMFP (10)	kg PM10-eq	4.57×10^{-1}	5.85×10^{-1}	8.13×10^{-1}
	SOx	G	8.30×10^2	9.60×10^2	1.20×10^3
	NOx	G	1.00×10^2	1.10×10^2	1.20×10^2
	PM10	G	8.03×10^1	1.13×10^2	1.45×10^2
Photochemical ozone	POFP	kgNMVOC	6.41×10^{-1}	6.80×10^{-1}	9.91×10^{-1}

(All abbreviations are defined in the reference [8.44].)

Aichberger, Christian, and Gerfried Jungmeier. 2020. "Environmental Life Cycle Impacts of Automotive Batteries Based on a Literature Review" Energies 13, no. 23: 6345. https://doi.org/10.3390/en13236345.

Powertrain

When it comes to the powertrain, there is less difference in mass between the different technologies than there is for other aspects of the vehicle. For an ICEV, the engine and associated transmission are in the range of 100 to 200 kg (220–440 lb) depending on the vehicle, the size of the engine, and whether it is front-, rear-, or four-wheel drive [8.45]. For a fuel-cell system, the mass of the fuel cell, electric motors, and small battery is of the order of 250 kg (550 lb) [8.21]. For a BEV, the mass of the electric motor(s) and associated inverters and power electronics is again of the order of 100 to 200 kg [8.46].

Given their similar masses and relatively similar materials, it is not expected that the LCA environmental footprint, excluding the in-use phase, of these different powertrains will be substantially different, at least relative to the footprints of the energy storage and the in-use phase.

For non-carbon dioxide pollutants, again it is not expected that there will be a significant difference caused by powertrain type on HTP, MDP, FETP, and TETP. However, it must be noted that copper carries higher impacts on these categories than steel or aluminum, so vehicles with higher levels of electrification in their powertrains (hybrids and BEVs), which have higher levels of copper in them, will have a higher impact here [8.47]. There are also indications that the catalysts and other parts of fuel cells have a higher impact on these categories, which along with their higher copper content will be significant [8.48].

Overall Trends

So, how can we draw all of this together? Clearly, there is a strong correlation between the mass of a vehicle and the LCA environmental footprint when we consider the individual parts of a vehicle. There is a rich seam of LCAs that look at the effect of vehicle lightweighting on the environmental impact of ICEVs ([8.49-8.53] are just a small fraction of the available resources). These studies generally show a fuel consumption improvement of 0.25 to 0.5 L per 100 km per 100 kg (220 lb) of mass reduction. Most of these reductions rely on the powertrain being downsized in proportion to the mass reduction, that is, that mass reduction does not lead to a performance improvement through a superior power-to-weight ratio.

For BEVs, Del Pero et al. looked at the energy use in the in-use phase and came up with a series of impact reduction values (IRV) (that is, kg carbon dioxide (equivalent) reduction per 100 km per 100 kg of BEV weight) under different average carbon dioxide intensities of European electricity grids [8.54]. The results are shown in Figure 8.8, where, as you might expect, more carbon dioxide-intensive grids, such as Poland's, show the greatest emissions reductions, and carbon-dioxide-light grids, such as Norway's, show almost no change.

FIGURE 8.8 GHG reductions based on reducing BEV mass from Del Pero et al. [8.54].

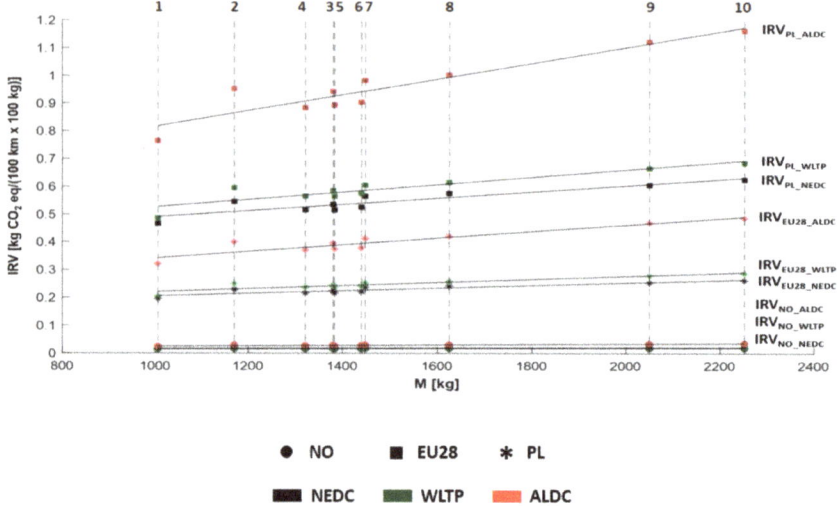

Del Pero, Francesco, Lorenzo Berzi, Andrea Antonacci, and Massimo Delogu. 2020. "Automotive Lightweight Design: Simulation Modeling of Mass-Related Consumption for Electric Vehicles." Machines 8, no. 3: 51. https://doi.org/10.3390/machines8030051.

Studies of the type that would answer the key question in this book—if you only have mass, can you get a good estimate of the total environmental impact of a car—are not yet available. However, Green NCAP has released a study that relates the mass of ICE, battery electric, and hybrid vehicles to their environmental impact. A net mass increase of 100 kg potentially results in an additional 500 to 650 kg carbon dioxide equivalent of greenhouse gas emissions across the different powertrain types in their analysis [8.55].

Another clue is that in a good proportion of LCA analyses, including in the GREET model and the Green NCAP study, many key carbon intensities are expressed as a function of vehicle mass [8.18]. Therefore, the environmental impact of a vehicle is seen as being tied closely to its mass in the operation of these models.

Summary

While throughout this chapter we have seen a strong correlation between vehicle mass and LCA environmental impact (across all categories, not just carbon dioxide), there is no perfectly clear direct link. That said, all the clues are there, in our opinion. Fundamentally, the more mass goes into a vehicle, the more environmental impact its production and disposal will have. That is simply because, for almost all materials, their environmental impact grows with how much of them you have. There will be some cases where switching to lightweight materials would increase the environmental impact of production and disposal of a car; however, the cost of such materials is typically so high as to act as a practical barrier to this for all but the most niche applications.

9

Non-exhaust Pollutants

IN THIS CHAPTER

- Cars create several pollutants when they are used that do not come from the tailpipe.
- Non-exhaust emissions from tire and brake wear are at least an order of magnitude higher than those from the tailpipe.
- Particles from tire wear are a major polluter, accounting for almost one-third of microplastic pollution.

We started by looking at the traditional view of vehicle emissions, which is centered around tailpipe emissions, and then moved to evaluating the upstream emissions in constructing the vehicle. Now, we must look at those emissions during vehicle use that do not come from the tailpipe. These have not historically attracted much interest—or, indeed, regulation—but they are now coming to the fore as combustion engines are much cleaner, as well as the growing penetration of tailpipe-free BEVs. This chapter will set out the nature of those emissions, assess the magnitude and potential environmental and health effects, and consider the relationship with vehicle mass.

BEVs are sold on the emotional image that they have no emissions: "ZEVs" (Figure 9.1). Cut off the exhaust pipe, and surely is this true? But, in reality, all vehicles emit. Yes, they emit different things, in different ways, at different rates, but there is, literally, not a single ZEV [9.1]. By terminating the exhaust pipe, what exactly are we losing? As with ambient air, the dominant substance in the exhaust is nitrogen, an inherently harmless gas. Equally harmless is some water vapor present. These aside, the most prevalent compound is carbon dioxide, the greenhouse gas, as we have seen in Chapter 3. The average mass of carbon dioxide emitted per second across a typical journey is approximately 3.1 g for European cars and 4.4 g for US cars. Put another way, 1 kg of carbon dioxide is emitted every 5.8 km in Europe and every 2.5 miles in the US [9.2]. In contrast, the average flow of carbon monoxide is 0.004 g/s (in both Europe and the US), and for nitrogen oxides it is 0.003 g/s and one-tenth of that in the US. Together, all the gases (including the completely harmless ones such as nitrogen that make up the majority) coming out of the tailpipe average approximately 10 g/s. This tells us that, setting aside the dominant and harmless nitrogen and water, tailpipe emissions on modern vehicles are predominantly carbon dioxide, with relatively trace concentrations of air pollutants.

FIGURE 9.1   Zero emissions, it is painted on the BEV!

CarlsPix/Shutterstock.com.

While we are not saying that trace amounts of these pollutants are harmless, understanding the masses involved will form an important context and backdrop as we look at non-exhaust pollutants. In the push for electrification, these non-exhaust emissions may be dismissed as small and irrelevant, or common to all vehicles; but how, in real-world operation, do the amounts of these non-exhaust pollutants actually compare to their tailpipe cousins? Bearing witness to the diversity and potential threat of these emissions, there are five main types of non-exhaust emissions: evaporative emissions, resuspension, brake wear, tire wear, and road wear.

We will go through each in turn and see how much of a problem they are and what factors determine their emissions levels in the real world. Before that, however, there is another element of traditional emissions thinking that must be discarded. When the environmental effects of exhaust emissions are thought about, the mind immediately goes to air quality. This is a term with no strict definition, but is normally used to describe ground-level pollutants that cause damage to humans and animals by direct inhalation, for example nitrogen oxides, or by how they react with other chemicals in the air to produce hazardous derivatives, for example how volatile organic compounds react to produce ground-level ozone, better known as a necessary, although not sufficient, ingredient in producing smog [9.3]. The emission of carbon dioxide from exhausts might be considered an air pollutant because of its contribution to planetary warming, but as the effects occur in the upper atmosphere and the inhalation of carbon dioxide at usual concentrations has no direct health effect for a human, most commonly it is not described as an air pollutant. Therefore, we need to think of environmental effects in terms of both air pollution and greenhouse gases, but that still neglects crucial areas.

As the tailpipe emits primarily gases, including water vapor, these fugitive products are not usually associated with soil and water contamination. As we begin to consider non-exhaust pollutants, these additional destinations become central. The study of non-exhaust pollutants is a burgeoning area, but only recently so, as a result of an academic paper by a number of academics led by Edward Kolodziej from the University of Tacoma in Washington State, US, which linked a single chemical (N^1-(4-Methylpentan-2-yl)-N^4-phenylbenzene-1,4-diamine) in rubber shed from tires with the mass die-off of coho salmon on the West Coast of the US [9.4]. The link between vehicle emissions and marine pollution was thus made. Researchers descended on the area in great numbers, investigating whether the same chemical—commonly called "6PPD" (thankfully), and which is a tire preservative—killed other fish species and organisms. From there, food crops were analyzed, which showed, for example, how 6PPD was being drawn up through the soil into the leaves of lettuce—this was published in a 2022 paper by researchers from the University of Vienna in Austria [9.5]. Chinese academics went further and found 6PPD and its derivative 6PPD-quinone omnipresent in human urine in another paper published in 2022 [9.6]. Therefore, when considering non-exhaust emissions, it is imperative to look at their effects on air, soil, and water and, through these media, on human and animal health.

It is, of course, true that exhaust emissions are not entirely gaseous, as one of the classic tailpipe pollutants is particulate matter, as discussed previously [9.7]. However, they are generally so well controlled now because of the filters in the exhaust (the DPFs and GPFs that we met in a previous chapter), that interest in them has rightly declined [9.8]. This may be premature, however. There are many vehicles on the road that predate the introduction of filters in Europe, and around half of gasoline vehicles still do not have a filter. In the US, this proportion is much higher, with few gasoline models being equipped with a filter [9.9]. Added to this, filters can break or be compromised or removed by deliberate action [9.10]. In this case, the levels of particulate matter emissions can rocket. Nevertheless, the total amount of larger particles is in significant decline. The remaining fine and nanoparticles are usually thought of as air pollutants as they stay suspended for a long time [9.3]. Nevertheless, they do eventually settle on soil or water, or—as has been observed—on the snow and ice of the Earth's poles [9.11]. Consequently, we should avoid the simplification of seeing tailpipe pollutants as just an air pollution problem, and recognize non-exhaust emissions as affecting air, soil, and water.

Evaporative Emissions

The first type of non-exhaust pollutant to consider is "evaporative emissions." Largely unknown outside of professional circles, these emissions are important but have been solved to a large, but not complete, extent [9.12]. Since the introduction of ICEs, fuel has evaporated from the tank and the fuel delivery system to the engine, especially in hot climates. This particularly affects gasoline, which is more volatile (easier to evaporate) than diesel [9.13]. There are approximately 150 different organic compounds in gasoline, including groups such as alkanes and aromatics, each of which has a different volatility, and propensity to react with other chemicals, to form particles in the air, and to add to smog [9.14]. The malign effects of this evaporating fuel were observed early in Los Angeles, California, and regulations to limit the rate of these emissions were introduced in 1989—relatively early in the regulatory process set by the federal 1970 Clean Air Act [9.15]. Activated carbon canisters were added to absorb expanding gas in the fuel line and then feed it back to the fuel tank. These canisters are highly effective, but there are multiple ways the hydrocarbons may still escape. Fuels may permeate the walls of the fuel tank, lines, and fittings. Vapor may be deliberately vented from the closed fuel system when pressure builds on a hot day. When refueling, even with the vapor recovery systems used in places such as California, some vapors escape. Finally, there are unintentional leaks through joints and other gaps—either leaks of liquid fuel or vapor [9.16].

The sum of all these routes for the fuel to escape to the air used to be a significant problem. With a small amount of unburnt fuel that can actually make its way through the engine and out of the tailpipe, in 2000, the average hydrocarbon emissions from gasoline and diesel vehicles in the US was 2.1 g/mile [9.17]. By 2021, this had been cut to 0.4 g/mile, an 81% reduction. Hydrocarbons in the air are undesirable primarily because of how they react—especially in hot and sunny places—to form smog and secondary particles [9.18]. By their nature, these emissions rates are primarily determined by the design of the fuel system, the specification of the activated carbon canister, and the ambient temperature, rather than vehicle mass. The fuel tank is likely to be bigger on a larger, heavier car, but this has a weak link to emissions rates.

Resuspension

The second type of non-exhaust emission is a complex, confusing, and hard-to-quantify one. Some would argue it should not even be considered an emission. As you drive along the road, some combination of the aerodynamics and the contact between the tire and the road stir up substances already lying on the road as "resuspended material" [9.19]. In other words, it is pollution whipped up by your car, but that did not originate from it. By stirring it up, there is no net addition to the total amount of pollution in the environment, and therefore, some say that it should be ignored as a non-exhaust emission. However, by resuspending the material in the air, in a sense it is reactivated. This reactivation means that it becomes available for human inhalation while it is in air. Even if that does not happen, the process of resuspension can help propel the material away from the road and onto the soil on neighboring fields or into nearby waterways—with all the potential effects on marine life and the food chain [9.20].

While this makes conceptual sense, what actually is this material? The reality is that there is no helpful definition to allow its quantification, and, by common understanding, it is highly diverse in its nature. In the mix will undoubtedly be finer particles from tires, brakes, and exhausts of vehicles, with potentially similar particles from industrial sources and home heating; coarser substances perhaps best described as dust and debris from neighboring farms, fields, and embankments; sand if you are in or close to a desert (Figure 9.2); and salt if you are close to the sea or it is winter and it has been spread on the road [9.19]. The last group is likely to have little environmental or health effect, even if inhaled by a human. These coarser particles tend to be caught effectively by the membranes in the nose and mucus high in the respiratory system [9.21]. The finer particles pose a similar danger as described in the chapter on tailpipe emissions (mostly their original sources). To make it more interesting still, these particles sit on the carriageway for a period, get stirred up by aerodynamics or tire shear, resettle, get stirred up, and so on. In this process, the particles can pick all sorts of volatile substances that might land and stick to their surface (adsorption in the jargon) or adhere during the mixing process [9.7]. For example, a gaseous exhaust pollutant might get mixed with and stick to a particle. This particle, if inhaled, would then deliver not just the particle but its harmful coating deep into the human body [9.22].

FIGURE 9.2   Resuspended dust.

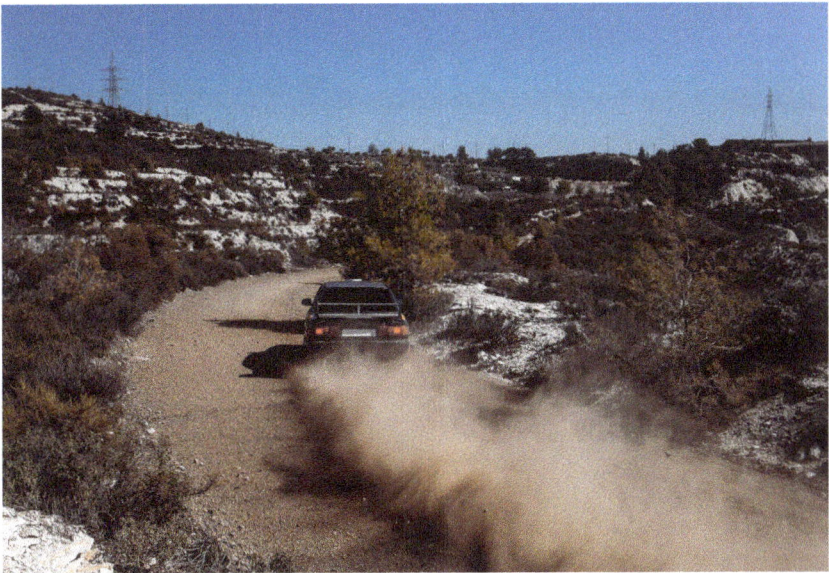

BoJack/Shutterstock.com.

Although we can describe these resuspended particles in this way, as there is no strict definition, any figures for the amount of resuspended material must be treated with caution. Many sources say the amounts are significant [9.5, 9.19, 9.23-9.25]. While this may sound concerning, there are many common situations where it is not surprising or particularly concerning. For example, if salt has been spread on the road during the winter, as it dries off, there can be significant resuspension of relatively harmless substances. A French study suggested that resuspension emissions might be three to seven times higher than exhaust emissions, but the potential harm is highly uncertain [9.26].

Even though the nature and rates of resuspended emissions are so variable, there is one element we can be relatively certain of: the main determining factor of the extent of these emissions is the mass of the vehicle. The speed will, naturally, also be a major determinant, but that is not an inherent property of the vehicle and so of no use to us in characterizing the environmental impact of a vehicle as a product. Although little academic research has so far been published that models the levels of resuspension and transit distances of the particles as a function of mass, it is intuitive that vehicle mass is the primary determinant. The heavier the vehicle, the greater the force between the tires and the road, which increases in a linear fashion. With the aerodynamic process of resuspended particles, the mass of the vehicle is not a factor. However, taking aerodynamic drag as a proxy for this effect, we know that the drag is a linear factor of the coefficient of drag C_d and the frontal area of the vehicle A_f (see Chapter 6). C_d is essentially independent of vehicle size; however, A_f will be larger for bigger vehicles (well, all practical vehicles), which correlates closely with vehicle mass. Therefore, although the connection between resuspended material and vehicle mass has not been explicitly studied, it is reasonable to believe that the amount will increase in at least a linear way with vehicle mass.

Brake Wear

The next non-exhaust emission to consider is brake wear. This moves us from the complex and unpredictable to something much better understood. Brake and tire wear, in a sense, are the only emissions that are "deliberate." With brakes, the friction between the pad and disc is necessary to slow the vehicle and wear almost inevitably results. These wear emissions are typically a combination of metals from the disc and resins from the pad, released in the form of fine particles [9.27]. The precise size varies with what the brakes are made of, but also particularly the brake temperature. Sizes in the range of 100 nm to 6 μm are typical (larger than GDI tailpipe emissions). However, at very high brake temperatures (>350°C—seen in hard braking) particles in the range of 11–29 nm (small even by GDI standards) are seen [9.28]. Brake discs are often primarily iron, with lower concentrations of copper, glass, and organic material. If released into the environment, metal particles such as copper can contaminate water systems and affect the growth and reproduction of aquatic creatures [9.29].

On the surface of it, the rate of brake wear should be closely related to the mass, and braking intensity, of the vehicle. The brakes must dissipate the vehicle's energy to slow it down. Most of this energy is turned to heat, with small amounts of noise. The amount of energy is in linear proportion to the mass of the vehicle. However, this is not how it works in reality, as the braking technology installed on the vehicle can have a much greater effect. Modern vehicles mostly have disc brakes, but on older ones (and some more modern, often cheaper, ones), drum brakes were preferred. These latter brakes pressed the pads into a rotating brake "drum," whereas the modern preferred technology is pads squeezed against an open disc using calipers. As a consequence of the geometry of the systems, disc brakes tend to create higher emissions, as the drums by their relatively enclosed design contain much of abraded material. Disc brakes have been preferred over the last generation because of their more even braking performance and because they do not overheat. There is, however, a movement back to drum brakes for electric vehicles, which primarily rely on regenerative braking, as the drums are cheaper and contain the abrasion particles more effectively [9.30].

A technology that is becoming an increasingly prominent feature of modern cars is "regenerative braking," and this also has a significant effect on brake emissions. Rather than converting the energy of the car to heat and expelling it to the surrounding air— quite a waste—the regenerative braking systems aim to capture that energy and store it in an onboard battery [9.31]. That electricity can then be used for purposes from powering auxiliary systems such as lights and entertainment systems—in mild hybrids, for example—or powering the car—in full hybrids and BEVs. Regenerative braking systems are an integral part of BEVs—this is simply running the motors as dynamos, thereby creating resistance. The systems also exist on some ICEVs and are sometimes referred to as "48 V systems" [9.1]. This is also a dynamo technology, which requires higher voltage electrical architecture than the standard 12 V, which has been the standard car battery voltage since the 1950s, to capture the energy and deliver it back quickly enough. The resistance created by the dynamo is what slows the vehicle and, as a useful side effect, reduces the need for the traditional brakes. The cars still need the traditional brakes for emergency braking, but the amount of friction braking is much reduced. As a result, two vehicles of the same mass, one with regenerative braking and one without, will have very different brake emissions—the former being much lower.

Research estimates of brake wear emissions from cars tend to fall in the 10–80 mg/km range [9.26]. This wide range may be due to encompassing both disc and drum brakes. Potentially half of these particles may become airborne and available for inhalation. Even at the low end of this range, brake emissions would be significantly higher than particle emissions from the tailpipe and higher even than many older cars. Therefore, these wear rates are significant in their quantum—and are predicted soon to become the single largest source of non-exhaust airborne pollutants according to CARB—but the weight-dependency of the emissions are dominated by other technical features of the vehicle [9.32]. To estimate the brake emissions of a particular vehicle, it would be more important to know the type of brakes and whether it has regenerative braking than the weight of the vehicle. That is not to say mass has no effect, but it is secondary.

Tire Wear

Competing with brake emissions to be the single largest source of non-exhaust pollutants are tire emissions. As a tire wears, or abrades, in normal operation, small [9.33] invisible particles are released—why else would your tires wear down (Figure 9.3)? Evidence is growing that these small particles are made up of some of multiple micrometers in diameters (2.5–25 μm), and a large number in the nanometer range (as low as 20 nm and typical peaks seen at 100 nm) [9.33]. This abrasion is caused by the acceleration, braking, and cornering forces between the tire and the road, as the tire slips and deforms [9.34]. We would naturally think that this wear comprises "rubber" particles, although only approximately 10% of a typical passenger car tire is natural rubber [9.35]. Most of the rest is synthetic rubber, developed during World War II to replace scarce natural rubber, and which is produced from crude oil. Due to this innovation, tires contain a smorgasbord of organic compounds, some volatile and others less so, and covering functional groups such as alkanes, alkenes, and polycyclic aromatic hydrocarbons, which can cause a range of human ailments from organ damage to cancer [9.35].

FIGURE 9.3   All the tread that was on this tire is now particulate matter in the environment.

LDVF/Shutterstock.com.

Emissions Analytics has tested almost 20 different sets of tires sold in Europe for their wear rates on the public road, by measuring the mass lost by the tires over thousands of miles. Figure 9.4 and Figure 9.5 show the patented sampling system for tire emissions testing. The average wear rate is 70 mg/km, and the range is from 38 up to 161 mg/km [9.36]. Early results from the same test of US tires are suggesting significantly higher rates, because of vehicles and their tires being larger on average and poorer quality roads. The relatively small range for the European tires—a factor of two between the highest and lowest—is noteworthy as the types of tires tested range from the cheapest to the most premium. This suggests that tires wear within a predictable range whatever the make.

FIGURE 9.4   Emissions Analytics' patented tire emission sampling system (side view).

© SAE International.

FIGURE 9.5 Emissions Analytics' patented tire emission sampling system (rear view).

Tire wear emissions are inherently caused by the weight of the vehicle—the mass of the vehicle and the gravitational acceleration rate. A hypothetically weightless vehicle would cause no tire emissions. It is, therefore, logical that wear emissions increase with vehicle mass—other things being equal. In controlled testing, Emissions Analytics found that tire emissions increased by 26% between a pair of comparable vehicles where one was 489 kg or 33% heavier [9.37]. From similar experiments, we find that the tire emissions increase roughly in proportion to the vehicle mass. But are other things generally equal? Heavier vehicles—most notably large SUVs and BEVs—tend also to require larger diameter and width tires to cope with the weight and—especially with full BEVs—the torque. Wider tires tend to exaggerate the wear further because of more "scrubbing," which often happens during low-speed cornering as the steering tire slips against the road surface [9.38]. Driving style also does affect wear rates [9.39]. If the driver makes full use of the torque on a BEV, the emissions may not just rise, but rise by orders of magnitude. In contrast, the most light-footed of eco-drivers can eke much lower tire emissions from an electric vehicle by making full use of the regenerative braking system. Although driving style exists to some extent in relation to the technology on the vehicle, it is not a simple, inherent characteristic of the vehicle, and indeed, the effect of driving style on tire emission can be both positive and negative. Therefore, we must consider this only as a secondary factor and not one directly relevant to our choice of the best metric to describe the environmental merits of a vehicle. There are also differences in the tire emissions rates between standard summer, "all-weather," and winter tires [9.40]. Winter tires have different tread patterns and contain different chemical compounds, which lead to good grip and low wear in the snow and slush, but wear fast on asphalt or concrete. Nevertheless, this is also a secondary factor to the weight dependency, which applies whatever the tire.

Thus far, we have described tire emissions in terms of the rate of loss of mass from the tires. The negative health effects of particle mass are well understood. However, there are a number of compounding effects as well. If we look at particle number as well, we can get an idea of the amount of ultrafine or nanoparticles. These are more available for human inhalation than larger particles and can cross into the bloodstream and the brain, as well as affect the respiratory system [9.7]. These ultrafine particles tend to be formed under the stresses of sharp braking and cornering of the vehicle. When measured in real time, sharp spikes in particle number coincide with these driving events. The heavier the vehicle, the more pronounced these spikes are. Whether larger or smaller particles, a further characteristic of tire particles is that, once settled in the environment or absorbed into the human body, a cocktail of organic compounds can leach over time [9.6]. Therefore, you have both the effect of the particles—most frequently irritating to lungs—and a covert release of potentially carcinogenic or mutagenic chemicals over time. A big, heavy vehicle is likely to lead not just to the simplistic rise in rubber mass loss, but the more hidden effects of ultrafine particles and chemical leachate.

Road Wear

The fifth and final type of non-exhaust emissions is road wear. We know it happens, but it is difficult to measure [9.41]. We know it happens because roads break up and need replacing—and that material has to go somewhere—and heavily trafficked asphalt shows indentations because of the weight of the vehicles passing. Setting up an experiment to quantify the rate of wear and to study the factors causing it, is hard as it is practically almost impossible to measure the mass lost from the road over time. If you try to collect particles on the vehicle or by the roadside as vehicles pass, what you collect is a mixture of road wear with material from tires, brakes, and other sources as well as resuspended road wear (which you would not want to double-count). Disentangling that is almost impossible, although some progress is now being made using chemical fingerprinting [9.42].

There is no consensus on the rate of road wear either, which reflects the lack of accepted methods and a small amount of research overall. The European Tyre & Rubber Manufacturers' Association (ETRMA) trade body has estimated that the rate of road wear caused by a vehicle is approximately the same as the tire wear on that same vehicle, although the nature of the method used is unclear [9.43]. Other quoted rates appear to track back to a small amount of very old research, of questionable accuracy or relevance to modern-day vehicles and road surfaces [9.44]. As a result, we have little idea of the true rates of road wear. However, we can be more certain about the dependences purely by logical deduction. The process of road wear is caused as a result of the force applied by the tires rolling over it. This is the only cause of wear. The force applied is a linear function of the mass of the vehicle, as we can recall from school physics, where the force is the product of mass and—in this case—gravitational acceleration, which is a constant. While we cannot know that the rate of road wear is in proportion to this force, it is reasonable to deduce that the wear rate is an increasing function with the force and, therefore, an increasing function with vehicle mass. There are no other material factors that will determine the wear. It is true that the exact design of the tire will have some effect, but that would be secondary. Other factors, such as rain on the carriageway are, like the speed of the vehicle, not an inherent property of the vehicle and so irrelevant to the mission of this book. Lastly, the nature of the road surface will also have an effect but is also not a feature of the vehicle. Asphalt and concrete are by far the most common surfaces. Each wears at a different rate—it is generally considered that asphalt wears faster, but even that material comes in many different forms and grades—but also the toxicity of the resulting wear particles will also be different [9.45]. Asphalt is potentially the worse of the two, as it is derived from crude oil and so is likely to have high proportions of aromatic and other heavier molecular weight compounds, some of which are carcinogenic and contribute to air quality problems when they evaporate off the surface [9.45].

Summary

Considering all five sources of non-exhaust pollutants together, we see that there are three—resuspension, tires, and road—that are strongly determined by vehicle mass. Evaporative emissions would, at best, have a weak link. Brake wear does, of course, depend on vehicle weight, but this factor is dominated by the design of the braking system, including whether it incorporates a regenerative braking system. To synthesize this together into an overall picture, we must consider the relative amounts.

Evaporative emissions capture systems work well such that only a small amount of fuel escapes, substantially reducing secondary particle formation in the air and other bad effects. Therefore, these will be a small part of the overall picture. The resuspension rate is highly variable and debatable, so it is difficult to put any value on that, and much of it by mass is dust and salt, which are less damaging to the environment and health. The average wear rate of brakes quoted about is 45 mg/km [9.27]. For tires, the equivalent rate is 70 mg/km [9.41]. We are guessing somewhat around the road wear, but if Michelin is right, that would add a further 70 mg/km, making a total of 185 mg/km for non-exhaust particle emissions by mass. Of that 185 mg/km, 115 mg/km—62%, or about two-thirds—is seen as strongly associated with vehicle mass.

But it is more significant than that. The maximum particle mass that is allowed from a current European passenger car is 4.5 mg/km, and in the US the equivalent is 3 mg/mile (1.86 mg/km). When the "Euro" emissions regulations were first introduced in Europe in 1992, the limit for diesels was 140 mg/km [9.46]. Therefore, the current levels of non-exhaust emissions are similar to the maximum permissible tailpipe particulates 30 years ago! Furthermore, the maximum tailpipe limit now is more than 80% below the actual non-exhaust levels (Figure 9.6). Emissions Analytics' research puts real-world tailpipe particles at approximately 0.02 mg/km, making them 99.99% below non-exhaust levels [9.47]—you can see this illustrated in Figure 9.7. Even though tailpipe emissions are more than just about particles—including the Dieselgate nitrogen oxides—we can now see through the perspective on non-exhaust emissions that most tailpipe emissions are now low, perhaps even scientifically irrelevant (see Figure 9.7 for an illustration of this in the UK), and that most non-exhaust emissions are a strong function of vehicle mass. That many of the tailpipe emissions are not a function of mass becomes much less relevant in this wider picture.

FIGURE 9.6   Comparison of non-exhaust emissions for three different car types and sizes.

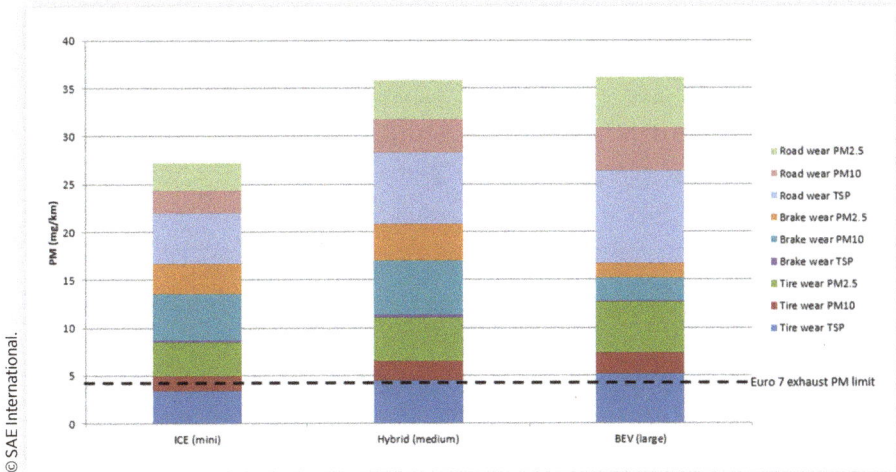

FIGURE 9.7   UK road transport PM$_{2.5}$ emissions showing the ever-increasing contribution of non-exhaust emissions [9.48].

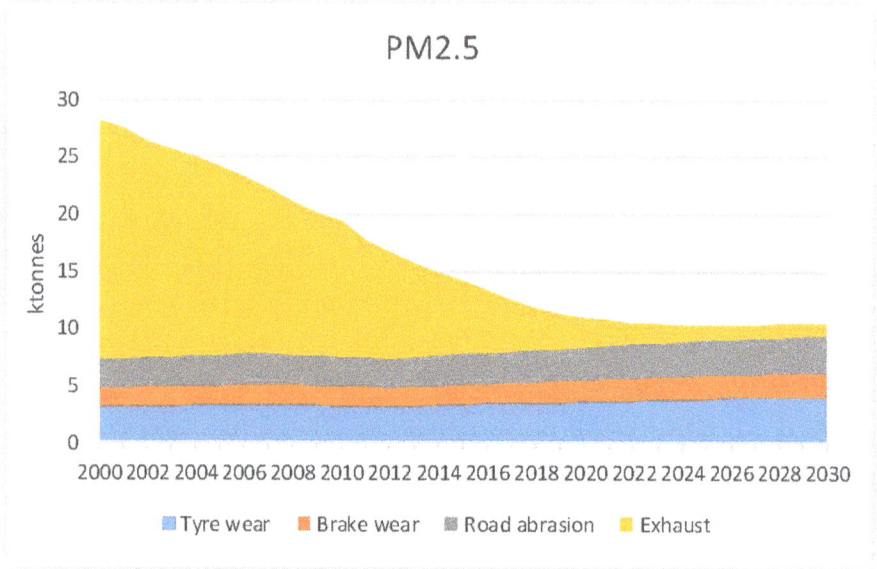

To avoid misinterpretation, despite the weight dependency of many non-exhaust emissions, they are not an exclusively BEV problem. In the specific case of brake emissions, these are likely to be less from electrified vehicles on average. More generally, one has to conclude from this analysis that non-exhaust emissions have been a greater problem for much longer than has generally been recognized. So caught up with exhaust emissions (which are easier to measure), we have neglected a source rapidly increasing its share of total emissions. In a sense, thinking is over a decade out of date. Prior to the widespread deployment of exhaust diesel particle filters, tailpipe emissions would have been dominant. Recall the puffs of acrid dark grey smoke from many vehicles at start-up and under acceleration (Figure 9.8). From 2010, this began rapidly to change, yet even now many students of air quality overestimate tailpipe emissions and underestimate the non-exhaust. This holds true even if—according to Emissions Analytics' estimates—only approximately 11% of tire emissions become airborne, compared to virtually 100% of exhaust particles [9.47].

FIGURE 9.8 Perception rather than reality today.

Juergen Faelchle/Shutterstock.com.

If there is any lingering doubt about the centrality of non-exhaust emissions in the environmental impact of vehicles, we can put them in the context of research on plastic pollution and, in particular, on microplastics. A 2019 report by Eunomia Consulting estimated that the total annual plastic release to the environment was approximately 12.2 million tonnes worldwide. Of this, 0.95 million tonnes were primarily microplastics, of which 0.27 million tonnes were from tires, or 28% [9.49]. We must conclude, therefore, that tires alone are a leading environmental contaminant. When other non-exhaust emissions more than double that amount, it becomes clear why tailpipe pollutants are relatively of declining importance in the holistic environmental impact of vehicles.

10

Noise, Safety, and Infrastructure

IN THIS CHAPTER

- Vehicle noise, safety and infrastructure are all adversely affected by vehicle weight.

- Safety effects are asymmetric: heavier vehicles are better for the occupant, but riskier for those outside that vehicle.

- Larger, heavier, vehicles have infiltrated, and affect, every part of the modern world.

The environmental impact of vehicles goes beyond even traditional tailpipe, manufacturing, and non-exhaust emissions, to include a range of "unconventional" impacts.

Noise

It is worth briefly thinking about why we are concerned with noise. It is not just about being able to listen to a favorite podcast in your car or sit in an idyllic pasture unbothered by a nearby road. It can be much more serious than that; noise pollution in urban environments is linked to increased risk of cardiovascular disease [10.1], low birth weight [10.2], and obesity [10.3] among other adverse health outcomes, notably through sleep disturbances [10.4]. In 2020, approximately 113 million people living in Europe were exposed to noise levels greater than 55 dB (the EU threshold for excess exposure defined in the Environmental Noise Directive), a major source of which was road traffic noise [10.5].

Vehicle noise comes from three main sources: aerodynamic noise, engine noise, and road/tire interface noise.

Aerodynamic noise is the din associated with the air moving around the body of the car and is generally reduced with a lower-drag shape. **Engine noise**—it is obvious where that comes from and is usually dependent on the engine speed and load, with the higher noise coming from more revolutions per minute and load on the engine. **Tire noise** comes from the interaction between the tires and the road and depends on the tire material, tire design (for example, tread pattern), and the road surface itself. For all three of these sources of noise, the greatest increase will come with increases in vehicle speed.

The impact, then, of vehicle mass on noise is not going to be clear-cut. There will be direct effects, but these might be masked by other non-mass-related things—primarily tire and vehicle design. However, in general, heavier vehicles will be larger and a larger vehicle will produce greater aerodynamic noise because it has a larger surface area. In addition, larger vehicles tend to have less aerodynamic (that is, higher drag) shapes and so will also produce greater noise.

Engine noise may be closely linked to vehicle mass for ICEVs. Heavier vehicles will usually have greater engine noise (there is simply a greater mass flow through the engine and hence higher noise). Of course, overly small engines for a given vehicle size will be working at a higher load consistently so will be noisier, and very high-revving engines in sports or race cars will also be noisy despite those vehicles being lightweight. Nevertheless, as a good approximation, a heavier vehicle will produce more engine noise. All of this said, engine noise is also a choice. Silencers fitted to engines, engine/gearbox matching, and tuning can also make a significant impact on engine noise. Some brands are renowned for their quiet driving experience—indeed a 2021 Rolls-Royce had to add noise to their 2470 kg Ghost model as the silence was making people uncomfortable (Figure 10.1) [10.6]. On the other hand, we all know of sports cars and motorbikes that create extremely loud noises despite being lightweight—many are designed in large part for their engine noise, such as the 1300 kg Volkswagen Golf VR6 (Figure 10.2). For fuel-cell vehicles and BEVs, the engine noise is negligible relative to the aerodynamic, road and tire noise.

FIGURE 10.1   Rolls-Royce Ghost—a silent and heavy experience.

TonyV3112/Shutterstock.com.

FIGURE 10.2   Golf VR6, small and very loud!

Stoqliq/Shutterstock.com.

Heavier vehicles often produce more **road/tire noise** because of the increased weight pressing down on the tires. This will lead to greater friction between the tires and the road surface, which will lead to increased road/tire noise levels. The mechanisms are similar to the tire and road wear mechanisms we discussed in the previous chapter.

Combining these three sources, experimental studies have shown that an increase in vehicle mass does lead to an increase in noise. Olson in 1972 [10.7] showed that a doubling of vehicle mass roughly leads to a 3.5 dB increase in noise—albeit with quite a large scatter—see Figure 10.3. Waters [10.8] found similar results in 1974 (see Figure 10.4). Also, Peng et al. [10.9] observed that heavier vehicles require a longer distance to accelerate, hence generating engine noise for a longer duration compared to lighter vehicles. Noise from heavier vehicles will also be increased going uphill as the engine load will be even higher under those conditions.

FIGURE 10.3   Data from Olson showing vehicle noise as a function of vehicle mass [10.7].

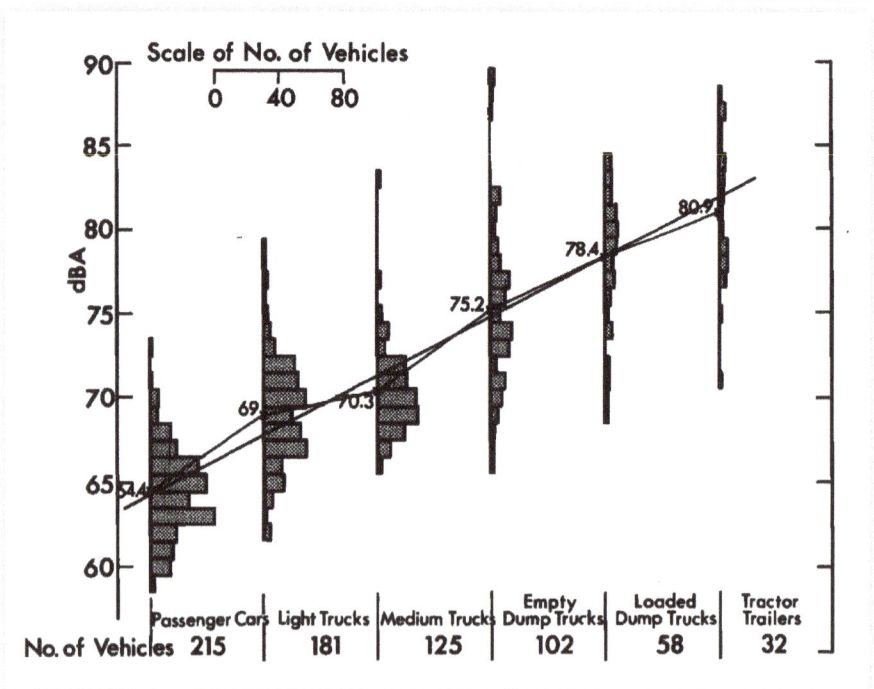

FIGURE 10.4 Data from Waters showing the influence of vehicle weight on vehicle noise [10.8].

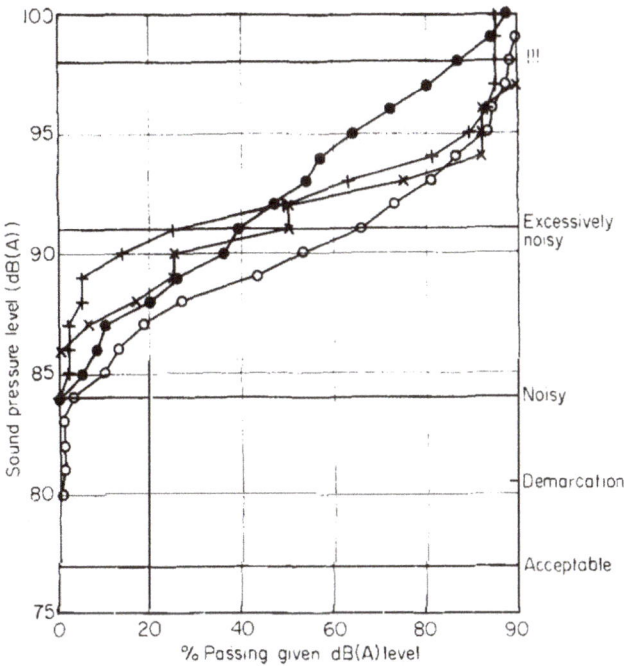

Overall, then, road noise will tend to increase with vehicle mass, but the drivers of this are indirect. There are examples of vehicles, particularly sports cars and motorcycles, which do not follow this trend, but even then, within those subcategories, these trends will be followed.

Safety

Automotive safety is an acute issue; approximately 1.35 million people each year are killed in car crashes [10.10]. The relationship between vehicle mass and safety is surprisingly simple. It is a story of two halves, however. For occupants of the crashing vehicle, a heavier vehicle will tend to have higher safety because, due to its increased mass, it will be able to withstand a greater force during a collision. This is likely to cause less injury to its occupants. Studies bear this out; for example, Klein et al. [10.11] in 1993 noted that a reduction in vehicle weight from 3700 to 2700 lb (1700–1200 kg) was estimated to increase the driver serious injury rate by 4% in Maryland, US, and 14% in Texas, US. Similarly, Høye [10.12] in 2019 in Norway found that an increase in vehicle weight increased the safety of the occupants of that vehicle by 39–54% even when adjusted for safety improvements of vehicles over the years (see Figure 10.6). Wenzel and Ross (2008, [10.13]) also find a weak trend between safety and vehicle mass (shown in Figure 10.5), although they do observe that differences between manufacturers were greater than differences because of vehicle mass.

FIGURE 10.5 Data from Wenzel and Ross showing the influence of vehicle weight on driver safety [10.13].

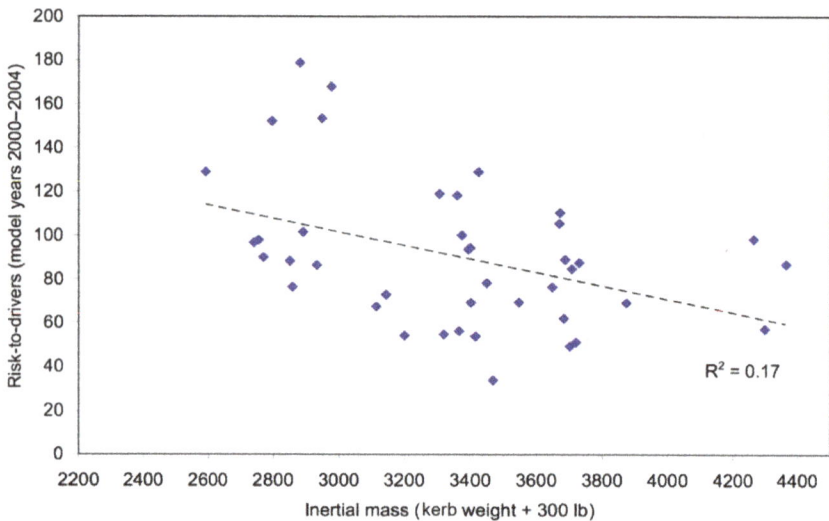

Reprinted with permission from "The Relationship between Vehicle Weight/Size and Safety," Wenzel, T. and M. Ross, AIP Conference Proceedings, 1044(1), 2008. Permission granted by AIP Publishing through Copyright Clearance Center.

On the other hand, for those who might interact with a vehicle, a higher vehicle mass means that there is more energy to dissipate in a crash, and therefore, the risks are disproportionately higher for those interacting with a vehicle—for example, pedestrians, cyclists, occupants of other vehicles—than they are for those in the crashing vehicle. Studies bear this out too [10.14, 10.15]; Simms and Wood [10.16] found that 11.5% of pedestrians struck by SUVs are killed, compared to 4.5% of pedestrians struck by passenger cars in the US; the study by Høye [10.12], which we just looked at for occupants of the vehicle, showed that for those outside the vehicle an increase in vehicle weight decreased the safety of occupants of that vehicle by 42–67% even when adjusted for safety improvements of vehicles over the years—see Figure 10.6. Tyndall estimated that every 100 kg increase in average vehicle weight was associated with an additional 0.03 fatalities per 100,000 people [10.17] and that a pedestrian is 70% more likely to die if the involved vehicle is a pickup truck rather than a car, and death is twice as likely if the vehicle is a large SUV rather than a car [10.18]. Edwards and Leonard similarly found that children are eight times more likely to die when struck by an SUV compared to being struck by a passenger car [10.19]. The evidence is overwhelming that heavier vehicles are much more dangerous to those who may be hit in almost all circumstances.

FIGURE 10.6 Influence of vehicle weight on safety from Høye [10.12].

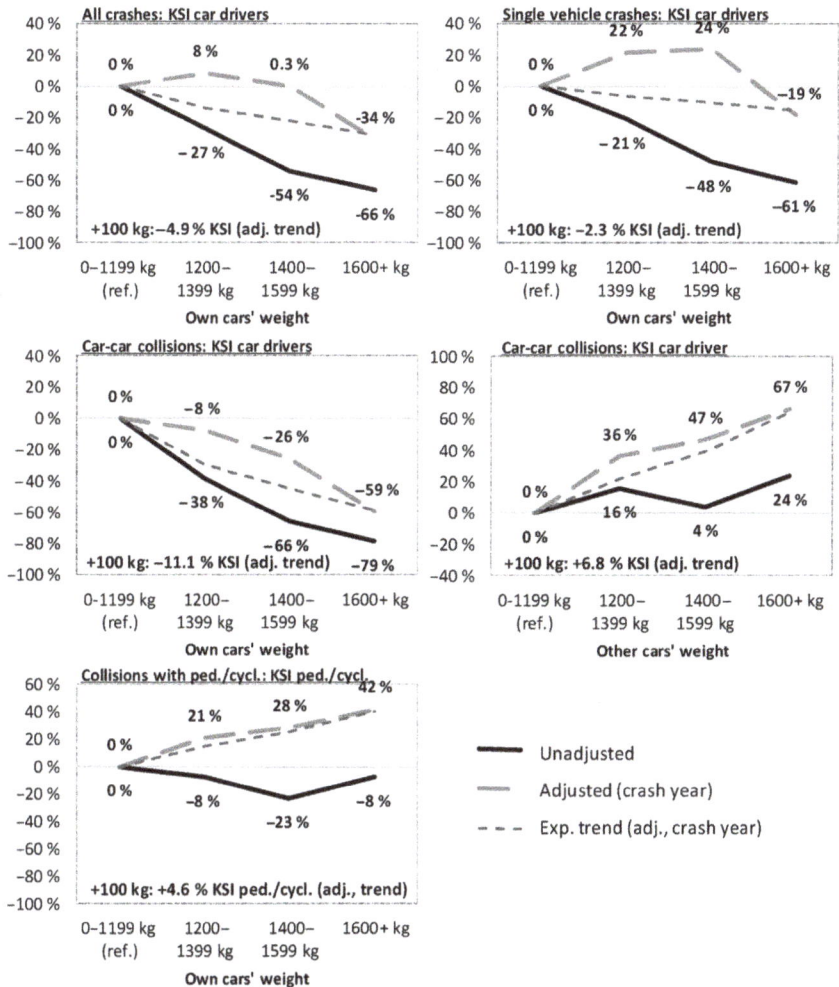

Evans and Frick in two separate papers [10.20, 10.21] investigated a number of different collision scenarios and looked at the mass ratio between the two vehicles. In one study [10.20], they also separated the effects of mass and size. Their results are shown in Figure 10.7, with an extremely nonlinear impact of mass to the ratio of driver fatalities (that is, whether the fatality was in the heavier or the lighter vehicle). They determined a relationship of $R = \mu^{3.5 \leftrightarrow 3.8}$, where R is the fatality ratio and μ is the mass ratio of the vehicles. In other words, the risk of fatality in the lighter vehicle goes up with more than the mass ratio cubed. What does this mean in reality?

FIGURE 10.7   Relationship between mass ratio of colliding vehicles and driver fatalities in a lighter vehicle from two different studies [10.20, 10.21].

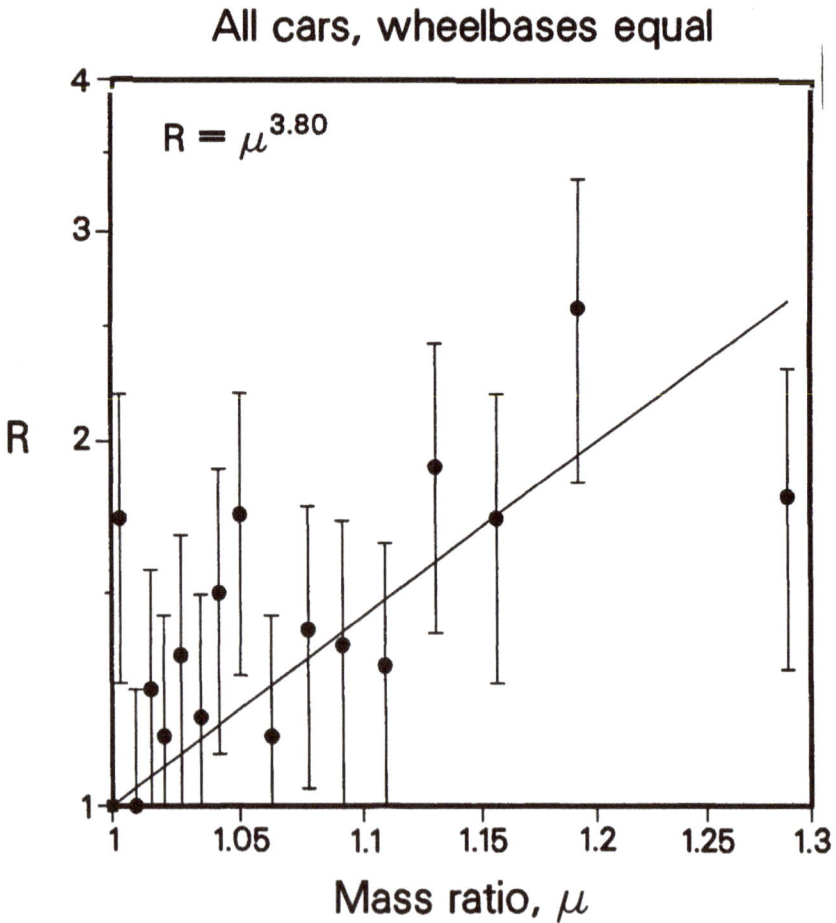

Reprinted from Car Size or Car Mass: Which Has Greater Influence on Fatality Risk? Evans, L. and M.C. Frick, Am J Public Health, 1992. 82(8): p. 1105-12 doi: 10.2105/ajph.82.8.1105.

(Continued)

FIGURE 10.7 (Continued) Relationship between mass ratio of colliding vehicles and driver fatalities in a lighter vehicle from two different studies [10.20, 10.21].

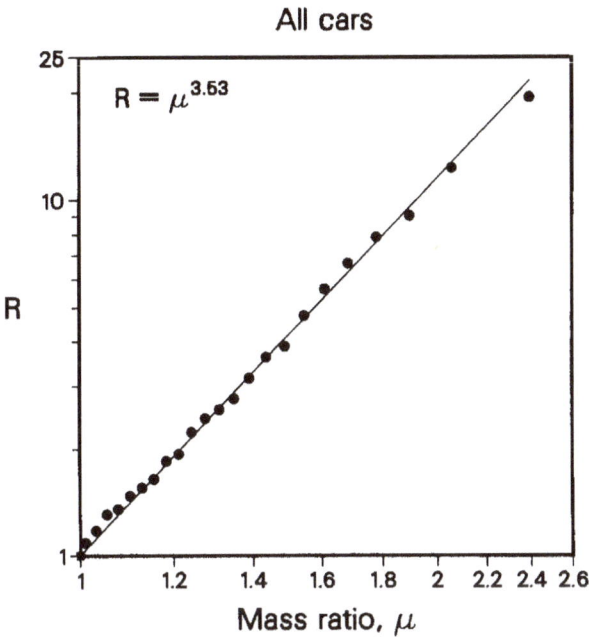

If you drive an average car (1500 kg in Europe or 1860 kg (4094 lb) in the US) and get hit by a car that is 100 kg (220 lb) more than the average, your risk of fatality goes up by 27% in Europe and 22% in the US. If you get hit by a car that is 500 kg (1100 lb) more than the average, your risk of fatality goes up by 198% in Europe and 147% in the US. In a later study, Padmanaban [10.22] reviewed the academic literature and concluded that vehicle mass was the most important contributor to the odds of a fatality in a collision, with a contribution of approximately 20–30% to variation in fatality odds. They also noted that in determining the odds of a fatality the next most important contributing factor was whether the striking vehicle in a collision was a light truck or not.

Heavier vehicles are also more likely to have an accident in the first place, because they may require longer stopping distances and have more difficulty maneuvering or avoiding collisions quickly [10.23]. This could potentially increase the risk of accidents, especially in situations requiring sudden braking or evasive maneuvers. This is particularly true for vehicles with a higher center of mass such as SUVs and pickup trucks. If these are maneuvered at high speeds, there is a substantial toppling and rolling risk, which makes these vehicles less safe.

Advances in materials, vehicle design, and crash technology continue to improve vehicle safety across the board, making it eminently possible for lighter vehicles to achieve high safety ratings, while heavier vehicles are not inherently safer in all situations.

Ultimately, while vehicle weight can be a factor in safety, it is essential to consider a wide array of elements, including vehicle design, technology, driver behavior, road conditions, and regulations to fully understand and improve the overall vehicle safety.

Nevertheless, the data is clear that the trends are that for the occupant of a vehicle, safety improves with increasing vehicle mass, and, for something that the vehicle hits, safety decreases even more with increasing vehicle mass. This is a classic example of an externality that our proposal will go some way to correcting. It is not an externality in the classic environmental sense, but it is an externality nevertheless, as a cost—in this case, a higher safety risk—is being placed on others by a vehicle driver that the driver is not bearing. It is possible to argue that the vehicle insurance system, to some extent, makes the driver pay for those extra claims, but that mechanism is uncertain and indirect.

Infrastructure

Since the 1950s, the impact of vehicle weight on road wear has been understood and investigated. We looked at this from an emissions perspective in the previous chapter. Here, we are interested in the impact on the road itself and its need for maintenance. For a long time, the relationship has been characterized by the so-called fourth power law, which says that the stress (and hence wear) on the road goes up with the mathematical fourth power of the axle weight of the vehicle. This is a widely cited "formula" which, among other applications, is used for road usage charges by trucks in the US, but we will see shortly that its scientific "law" status is not totally the case. In the US, the Federal Bridge Gross Weight Formula, also known as Bridge Formula B or more usually as the Federal Bridge Formula, was passed into law by the Federal-Aid Highway Act Amendments in 1974. This formula is used to determine maximum vehicle weights for trucks in order to avoid damage to roads and bridges. Therefore, we know about the strong effect of vehicle weight on infrastructure damage, and it has been legislated for over 50 years.

However, the general applicability of the fourth power law was really only designed to apply to overall deterioration of a road surface rather than wear or failure in a specific place. Its applicability has also been widely discussed [10.24]. For example, it only considers vehicle weight and not volume of traffic, and it is known that the wear from two 5-tonne vehicles is greater than from one 10-tonne vehicle. Rhodes et al. [10.25] investigated this and came up with a relationship between the power law exponent and the amount of traffic, shown in Figure 10.8.

FIGURE 10.8   Influence of traffic volume on power law exponent from Rhodes et al. [10.25].

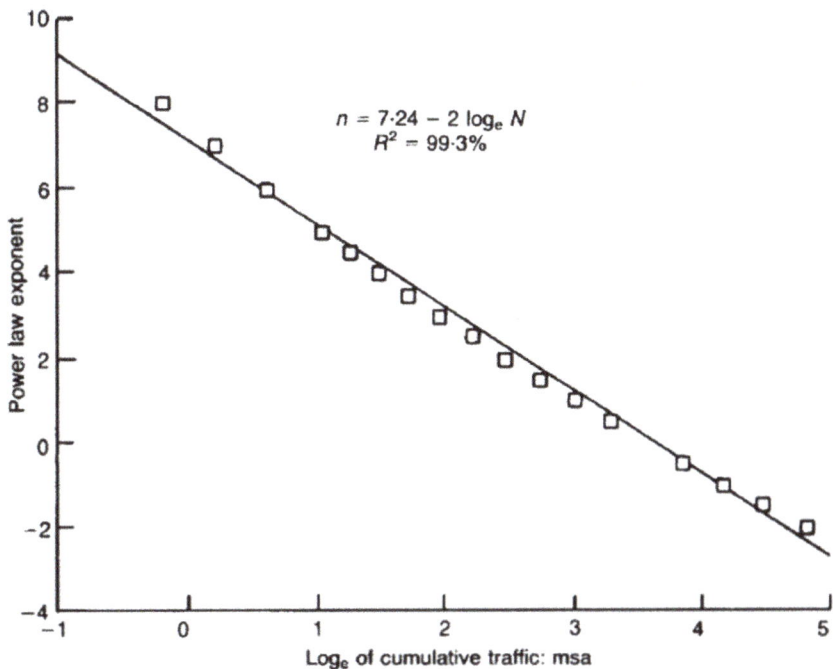

$$n = 7.24 - 2 \log_e N$$
$$R^2 = 99.3\%$$

x-axis: $\log_e$ of cumulative traffic: msa

y-axis: Power law exponent

It should also be noted that the fourth power law describes a dependency on axle weight, not vehicle weight. This is, of course, why trucks and buses often have more than two axles to distribute the weight better and, at least in part, reduce road wear. You may have seen road signs such as in Figure 10.9, an example from Ohio, US, where the total weight that can pass over a bridge or other part of the road system depends on the number of axles on the vehicle. Nevertheless, practically all cars (the main focus of this book), motorbikes, and many buses and trucks have two axles, so the axle weight is going to be directly proportional to vehicle weight for all but a few extreme examples.

Road wear is determined by the stress, either compressive or lateral, that the road surface experiences. Stress is defined as force divided by area and is similar to pressure—and has the same units as pressure. Therefore, efforts to "spread the load," for example, by bigger or more tires will reduce road wear. So will reducing the pressure inside vehicle tires, which increases the contact surface between the car and road, and hence decreases the pressure/stress on the road. The latter, of course, will increase the energy or fuel consumption of the vehicle as well as emissions from the tires, as discussed in the previous chapter.

Vehicle weight is not the only factor to affect road wear, as it also goes up drastically with vehicle speed as well as during high lateral load caused by vehicle acceleration or cornering [10.26]. You can often see this at bus stops, where a bus will often turn its wheels while stationary (lateral load) and then accelerate away from the bus stop (increased lateral load)—see Figure 10.10. Of course, it is not all about the vehicles that travel over the roads, as road wear is also affected by temperature. In very hot weather, tarmac can melt, and in winter, potholes are often caused by repeated freezing and thawing of water, which expands as it freezes.

FIGURE 10.9 Ohio bridge weight limit sign showing different total vehicle weights permissible for different numbers of vehicle axles.

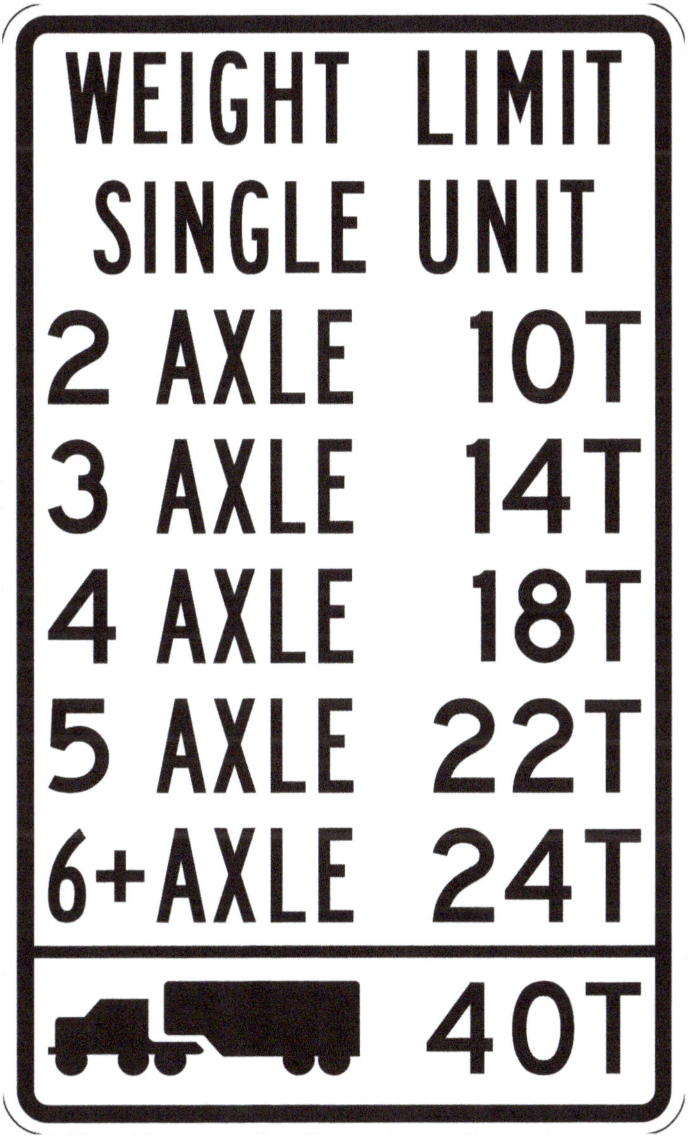

Reprinted from the Ohio Manual of Uniform Traffic Control Devices.

FIGURE 10.10   Road wear at a bus stop—here you can see multiple road resurfacings have taken place and the characteristic double depressions from the bus wheels are still present.

Road wear is not the only infrastructure issue that arises with increasing vehicle mass. Bridges have a certain design load, plus margins for error. Many bridges are decades, or even centuries, old and may not even have been designed for cars. As vehicle masses increase, the bending moments and shear forces that the bridges experience also increase. Either of those will have an upper limit beyond which the bridge will begin to fail. Regular loading and unloading can also induce fatigue-related failure, so increasing numbers of heavier vehicles has a compounding impact.

A notable example of bridge failure because of increasing numbers of heavier vehicles is Hammersmith Bridge in west London in the UK. The bridge opened in 1887 and was substantially refurbished in 1973. However, by 1997 it needed closing for a year for strengthening and was further strengthened in 2014 and 2016. The bridge has been permanently closed since 2019 (Figure 10.10) because substantial cracks had appeared in the structure, and it is not anticipated to reopen until 2027 at the earliest. So serious was the damage to the bridge that even the River Thames underneath was closed.

FIGURE 10.11 Hammersmith Bridge in London—a vision of our future?

This problem is not just restricted to a single bridge in London, as bridges across the UK are being inspected to check their structural integrity in the face of ever-increasing vehicle weight [10.27]. In the US, increased weight on the bridge was, in part, linked to the collapse of the I-35W Mississippi River Bridge in Minneapolis, Minnesota, which killed 13 people. In the US, approximately 150 bridges per year—mostly smaller bridges in rural areas—collapse because of their poor condition and heavy vehicle usage [10.28].

So what, you may think—old bridges and roads need repairing and proper maintenance, this is nothing new! Yes, that is true, and it is a long-running problem; the American Society of Civil Engineers estimated that there was a $786 billion backlog of roadway system maintenance caused by years of underfunding [10.29].

In the UK, the maintenance backlog is estimated at £20 billion ($25 billion), with the average road now being resurfaced every 116 years [10.30].

The poor condition of the wider transport infrastructure is also not independent of vehicle weight. Indeed, increasing vehicle weight exacerbates the problem. The UK Institution of Structural Engineers released new guidance for car park design in 2023 following a 33% increase in vehicle weight [10.31]. This design guidance is not just for new car parks, but specifically refers to maintenance and upgrade of existing car parks as well. This is not just alarmism or opportunism from structural engineers—they have never struck us as a particularly alarmist breed. In April 2023, a parking garage collapsed in New York City killing one person (Figure 10.12). In the aftermath, several other parking garages in New York City were closed after inspections because they posed an immediate threat to public safety [10.32]. At the time of writing, the investigation is still underway, but initial reports indicate that the cause was due to excess weight from more than 50 cars (of unspecified types) parked on the roof of the parking structure [10.33].

However, fixing infrastructure takes significant time and money, and it will take a substantial investment to put these things right. Heavier vehicles (whether electric or otherwise) are already on our roads today, and their numbers are rapidly increasing—we will look at how the average weights of cars are growing worldwide in the next chapter. The industry is aware of the looming problem and, in the US, the industry has been lobbying lawmakers to increase weight limits on roads [10.34]. However, weight limits are in place for an engineering rather than political reason. Safely changing weight limits will need substantial inspection of infrastructure as well as reduced weight limits in some places, and significant funding for upgrading roads, bridges, car parks, and other infrastructure worldwide. Fundamentally, this is another externality. Here, there is a physical danger to others from infrastructure damage or collapse and the additional repair cost will be covered by some combination of private property owners, their insurers, and taxpayers. The owners of heavier vehicles, who bear responsibility, do not pay.

FIGURE 10.12   Aftermath of April 2023 New York City parking garage collapse.

City Design

It is not just about weight; size matters too. Worldwide, most cities were not built for large cars, certainly not for SUVs and large pickup trucks. Many cities, particularly in Europe, were designed before the horse and cart, let alone the motor car. Even more modern cities were not designed with vehicles the size we have today in mind—they were designed to handle cars from decades ago. This has impacts on everything, from the size of parking spaces and driveways, to the streets themselves. Are size and weight perfectly correlated? No, but as we have seen previously in this book, they are well correlated, and so if we incentivize lighter vehicles, we are likely to be incentivizing smaller vehicles too.

This might initially seem like a trivial point, but this matters. When our cars get big, everything needs to get bigger to accommodate. Bigger driveways, wider streets, more sprawling parking lots. This costs money not just to build them but also for the land on which they sit. Land is expensive in densely populated areas. Who pays for this? People expect to park their car near or outside their house, but often expect to do so for free—ultimately storing private property on public land at no or low cost. To illustrate this, Felix's car has a footprint of 7.57 m^2, and a parking permit where he lives in Oxford costs £70 ($89) per year. This is £9.24 ($11.74) per m^2. Near the area of Oxford where Felix lives, office rental is available at £452 ($574) per m^2 (using data from Zoopla), making £3422 ($4350) for the footprint of the car. Equivalent numbers for London are £155 ($197) per car per year and £18,000 ($22,900) to rent the equivalent office space [10.10].

Certainly, cities in the UK have a tension at the moment around how we allocate space. Do we prioritize cycles, pedestrians, car parking, car running, or buses? What about loading for trucks and delivery vans? No city has the space for all of these things, and we just try and squeeze more in. Scrapes increase and tempers fray as a consequence. For an illustration of how car size has really exacerbated this, consider Figure 10.13. A typical street designed in the late 19th Century had plenty of room for parked carriages on both sides and two running carriages in the middle—we are using the size of the 1894 Daimler Motorized Carriage as an example.

By 1975, we were probably down to one running vehicle—here the 1975 VW Golf Mk 1 is our example. Today, we can barely squeeze a single running vehicle through—in this case a 2024 Jeep Grand Wagoneer. You have probably experienced this in cities near you already, and it is likely to get worse.

FIGURE 10.13　How what is on our streets has changed.

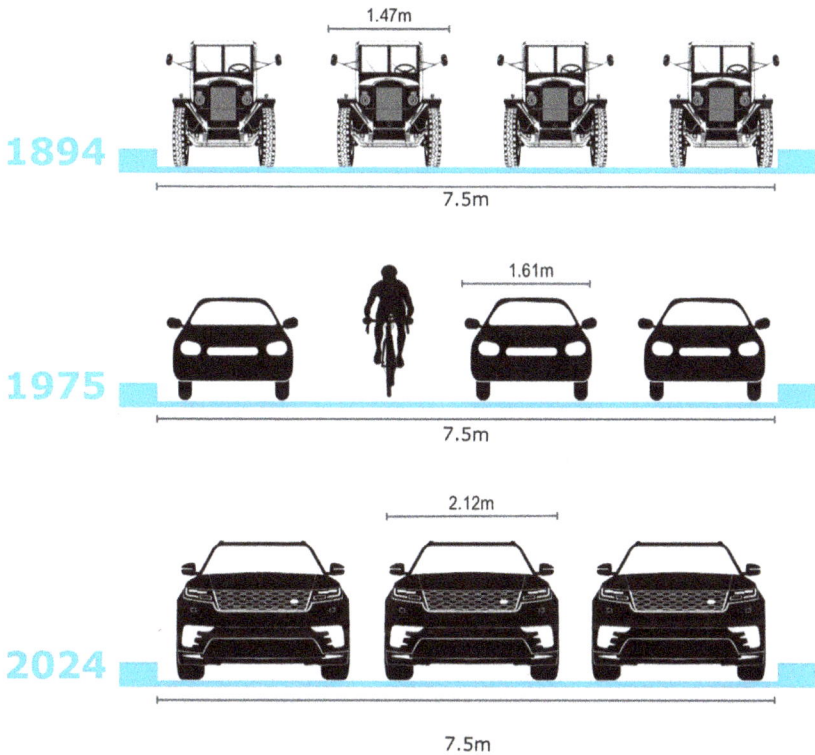

It is not just in cities. Other pieces of infrastructure were built for smaller vehicles; for example, ferries have a fixed amount of space for vehicles on board—if cars get bigger, the carrying capacity of the ferry (in terms of the number of cars) must reduce. See Figure 10.14 for an example of this in reality. Similarly, houses with smaller garages or driveways might not fit modern vehicles, perhaps pushing these vehicles onto public land or preventing updating your vehicle to a newer model.

FIGURE 10.14 Cannot squeeze into that gap!

We are also normalizing these changes and vehicle growth. Lego, the popular children's set of toys, has been widening its roads since the 1990s—keeping up with the growth in the real world (Figure 10.15). The Lego pavement (sidewalk) has shrunk from nine studs wide in the 1980s to only six today, with all of that extra space going to (toy) cars. Correspondingly, roads to drive them on have grown from 14 studs wide to 20 studs wide [10.35]. Of course, Lego is just keeping up with observations of the world; children love to play with accurate representations of what they observe. However, if children grow up with this as normal, any sort of reversal becomes even more challenging.

FIGURE 10.15 Even Lego roads have needed widening!

1980s   1990s   2000s

© SAE International.

Summary

So, there we have it. Vehicle weight plays roles in noise (indirectly), safety (very directly), and our infrastructure (directly)—even our toys (through size). Almost always, the increase in costs borne as a result of these negative effects is not paid by the owners or operators of the vehicle themselves. It is an externality paid by others and often us as taxpayers. If we continue to allow these heavier vehicles on our roads without some sort of penalty, we will all pay the price.

11

Beware What You Wish For

IN THIS CHAPTER

- Cars around the world are getting heavier in every market segment, notably in the last 15 years.

- Individual car models become heavier with each new generation, but consumers are also tending to upgrade to larger models.

- Regulation and legislation are often making automotive obesity worse.

The automotive sector has always been shaped by regulation, in part because it is an industry that generates many externalities. As discussed previously, externalities are a particular form of "pollution" where the person creating it does not fully pay for or suffer the consequences of it. Just like if you drive an old banger or clunker, it may be cheap travel for you, but it puts out high emissions that those around you have to suffer, at no cost to you.

Regulation and Legislation

These externalities are most often dealt with by legislation and regulation; indeed, it may not be taking it too far to say that legislation has been the biggest driver of innovation in the automotive industry in recent times [11.1]. Due to this, legislation is powerful in shaping the automotive industry and the vehicles on sale. Therefore, it is important that the regulations are well designed to achieve the desired goal. Sadly, this is not always the case, leading to undesirable side effects. On the consumer side, we might all as individuals choose a certain type of car that is right for us but neglect the downside if everyone made the same choice. This chapter will examine how the nuances of regulation and consumer preference are leading us to even bigger and heavier vehicles.

Cars have been regulated one way or another since the 1960s. Initially, the legislation came from California with the 1966–1967 Mulford-Carroll Air Resources Act, which was arguably the first ever emissions legislation. Around that time, legislation was also being developed worldwide on fuel quality with a particular focus on removing lead from gasoline, which was mostly accomplished by the year 2000 [11.2]. This first round of legislation was almost entirely around pollutant emissions, spurred by major smog events in big cities worldwide, including Los Angeles and London.

In the US in 1975, the first fuel economy standards (the Corporate Average Fuel Economy (CAFE) standards) appeared, spurred not by a desire to reduce carbon dioxide emissions, but because of the 1973 oil embargo and associated concerns about energy security. The successors to these 1975 CAFE standards are still in force today in the US [11.3].

Legislation is already adapting to, and in many places aggressively promoting, shifts in vehicle powertrain technology type, including toward BEVs. The UK has recently made legislative modifications to allow holders of standard driving licenses to drive heavier vehicles than were previously permitted, to make an allowance for the fact that BEVs are much heavier than their ICEV equivalents [11.4]. Similarly, in Europe, recent legislation applying to ICEVs around RDE compliance may now be extended from vehicles weighing up to 2610 kg to vehicles weighing up to 2840 kg at the manufacturer's request [11.5].

Another legislated aspect of vehicles is safety. We have already looked at the impact of vehicle mass on safety in the previous chapter. There are three main automotive safety standards that are used today: ISO 26262 [11.6], the Federal Motor Vehicle Safety

Standards (FMVSS) [11.7], and the New Car Assessment Programme (NCAP) [11.8]. ISO 26262 is an international standard focusing on electronic systems and software in vehicles, ensuring they meet certain safety requirements. The FMVSS is a wide-ranging set of US standards that cover aspects of vehicles from seat belts and airbags to flammability of materials and fuel system integrity. NCAP is a voluntary program that has different standards per country (there are 13 different NCAPs). These are focused on testing of vehicles under standardized conditions, as distinct from FMVSS, which sets out standards of construction. You have probably seen images or videos of these tests with crash-test dummies in the vehicles and high-speed cameras rolling. See Figure 11.1 for an illustration of this. NCAP also provides easy-to-understand information for consumers through a safety star rating for individual vehicles. A more detailed report is publicly available on their website for those who want greater detail.

FIGURE 11.1   A crash test.

Benoist/Shutterstock.com.

Most of these regulations and standards focus on the safety of occupants of the vehicle rather than anything, or anyone, the vehicle might hit, although some, such as NCAP, include pedestrian protection standards. We mentioned externalities previously, and we have already seen that there is a trade-off when it comes to vehicle size between safety for its occupants and others not in that vehicle.

These automotive regulations and their equivalents world-wide play a significant role in shaping the design, engineering, and market trends of cars. There is a whole chapter on the legislation regulating cars in Felix's previous book [11.9]. In some cases, these regulations have indirectly, or at least inadvertently, incentivized the production of larger and heavier cars because of various safety, emissions, and performance standards [11.10]. Hence the title of the chapter: Beware What You Wish For. If you create a regulation that embodies a certain incentive or disincentive, make sure you are comfortable with the likely outcome.

Fuel Consumption and Carbon Dioxide Emissions Standards, and Vehicle Size

When considering carbon dioxide emissions and fuel economy standards, there are many versions worldwide, some of which are, or have been, based on vehicle size and weight. Some of these, indirectly incentivize car manufacturers to produce larger vehicles that need to meet less stringent emissions targets. This seems counter-intuitive: Why might regulators want to make a set of standards that incentivize bigger and more polluting cars? The idea is that consumers want to buy a car of a given size to meet a given need and so, to reduce emissions from all cars, there should be some normalization relative to vehicle size. In addition, they are not starting from a "clean sheet"; vehicle manufacturers have existing product ranges and consumers have established preferences. This is called an "endowment effect," which is effectively grandfathering in certain biases.

Let us look at the current US CAFE standards as an example. These standards have been in force, in one form or another, since the 1970s and, as the name suggests, are applied to an average of a manufacturer's fleet and can be averaged over up to a five-year period [11.3]. This means that the average vehicle needs to meet these requirements rather than each individual vehicle. Manufacturers that do better than the requirements can sell "credits" to manufacturers that do not—creating a market for reducing fuel consumption and carbon dioxide emissions or, depending on your perspective, allowing people who pay to pollute.

The standards, at the time of writing, are shown in Figure 11.2 [11.11]. As you can see from the figure, over the years the required average miles per gallon (mpg) level has risen from under 20 in the 1970s to over 30 for light trucks and over 40 for cars in the 2020s. As carbon dioxide emissions come from the fuel a vehicle uses, this increase in fuel economy target is exactly equivalent to a decrease in carbon dioxide emissions, for the same type of fuel. Each liter of gasoline used in your car emits approximately 2.3 kg of carbon dioxide, and each liter of diesel emits approximately 2.7 kg [11.9]. As a guide, over the 1970s to 2020s, an increase in fuel economy from 20 to 40 mpg reduces carbon dioxide emissions from roughly 275 to 137 g/km[1] [11.12]. It is also worth noting that, by definition in the standards, vehicles that do not have a tailpipe (BEVs and FCEVs) do "infinite" mpg and have zero carbon dioxide emissions. This is what is really pushing, in particular for cars, ever higher mpg levels, which is expected to be particularly rapid in the later part of the 2020s with large numbers of these so-called ZEVs as part of manufacturers' fleets.

---

[1] Throughout this chapter, we acknowledge the useful conversions between carbon dioxide and mpg from the ICCT [11.12].

FIGURE 11.2   CAFE standards for vehicle mpg (US gallons) over time (data from [11.11]).

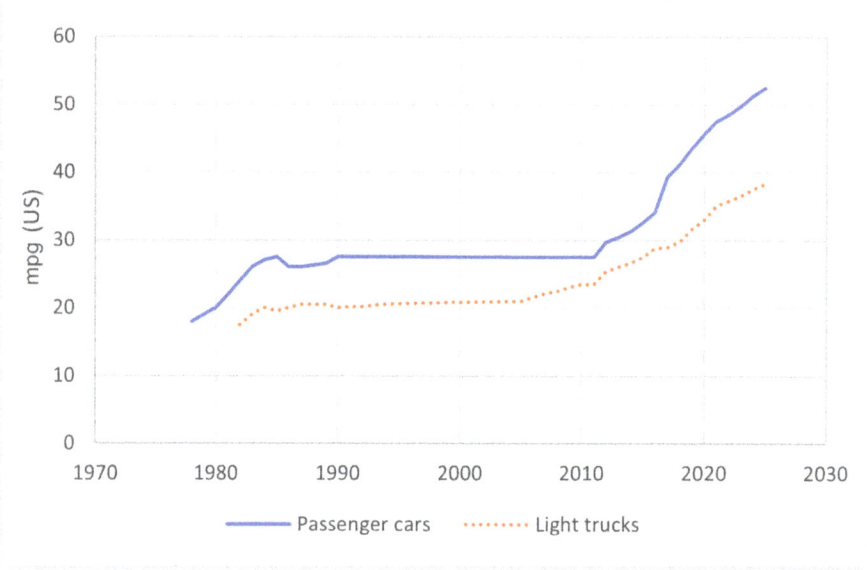

When thinking about these CAFE numbers in the context of our thesis, the first thing that jumps out is that there are different standards for cars and light trucks. The definition of a light truck is quite specific, but as a guide, think of a minivan, four-wheel drive SUV, or pickup truck [11.11]. As we have already seen, in one way this makes sense as bigger cars will have inherently higher fuel consumption, so this needs to be accommodated. However, if the objective is to reduce carbon dioxide emissions or reduce fuel consumption, then there is a chance that this approach will simply lead to more trucks and fewer cars. The mechanism for this is that the lower fuel economy target for light trucks makes the compliance cost lower and, therefore, the vehicles are cheaper to buy. As a result, some consumers will switch from buying a car to buying a truck. It should be noted that the Biden administration has recently proposed new CAFE targets through to 2031, which aim to reduce the gap between cars and light trucks.

There is more going on in the details of the regulation. Since 2010, vehicle footprint has been part of a normalization factor in the standards. Vehicle footprint is defined as the area enclosed by the four points where the tires touch the ground, that is, by multiplying the vehicle's wheelbase by its average track width. Vehicles with larger footprints, which are almost always heavier, get lower mpg and, therefore, higher carbon dioxide targets. On the surface of it, this is to be fair to manufacturers with product portfolios skewed toward larger vehicles. At the same time, however, it reduces the compliance cost for those larger vehicles, thereby reducing the retail price and increasing consumer quantity demanded.

The more cynical among you may also note that in the US foreign car manufacturers such as Honda or Toyota typically make smaller cars than the US "big three" (Ford, General Motors, and Stellantis, formerly Fiat Chrysler), and so this rule enables the big three to meet the standards more easily.

The precise way that this is implanted in the CAFE standards is as a mathematical function, which has evolved over the years [11.13, 11.14]. Figure 11.3 shows these functions for a range of model years, with carbon dioxide limits reducing over time. The figure also shows the relationship between vehicle footprint and the emissions limit, and demonstrates the greater permissiveness for larger vehicles within the same model year.

FIGURE 11.3 CAFE carbon dioxide standards with respect to vehicle size [11.13, 11.14].

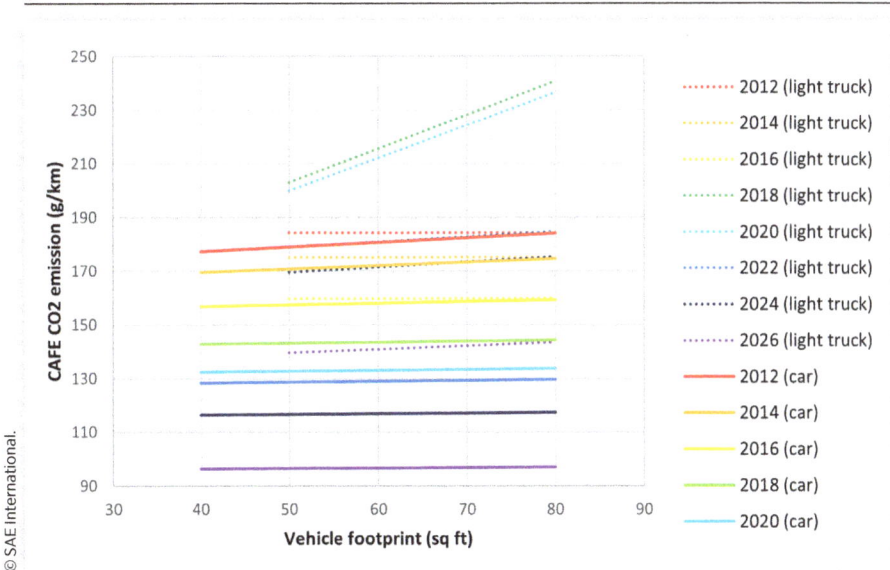

© SAE International.

What does this mean for some example vehicles? We have taken the footprints of a selection of vehicles and calculated their target CAFE mpg values. These are shown in Table 11.1. The Tesla Model X is an electric car, which by the CAFE definition does infinite mpg, but the value shown is what the minimum it "should" emit under the standards.

TABLE 11.1 CAFE mpg target values for some vehicles.

Vehicle	Designation	Vehicle footprint (ft²)	2024 CAFE mpg value
Volkswagen Up!	Car	42.7	47.2
Honda Civic	Car	53.0	47.1
Volvo XC90	Light Truck	54.0	32.2
Tesla Model X	Car	54.6	47.1
Ford F-150	Light Truck	77.8	31.4

© SAE International.

It is true that the CAFE standards have driven substantial increases in fuel efficiency—and hence reductions in carbon dioxide emissions over the years. However, the footprint normalization contains a second-order bias toward larger vehicles. As we noted previously, the theory behind the allowance for vehicle footprint was that certain consumers needed a vehicle of a certain size. However, the incentive for vehicles to get bigger was there, and indeed, this is what has happened [11.15, 11.16]. As we will see later on, consumer preferences have had a big influence too, and the makeup of the US and worldwide fleets is shifting rapidly not just to bigger footprint vehicles, but also to more light trucks and fewer cars. Indeed, Ford famously announced in 2018 that it would stop making passenger cars, although there would rightly be debate as to whether some pickups, SUVs, and crossovers would now be thought of as "cars" in common parlance [11.17].

We have seen that both safety legislation and fuel economy (and carbon dioxide emissions) standards can incentivize larger vehicles, but there is a tension here with other legislation. Historically, the US EPA has wanted consumers to buy vehicles with lower carbon dioxide emissions, while, at the same time, the US National Highway Traffic Safety Administration (NHTSA), another branch of the same government, has expressed concerns that smaller, lower carbon dioxide (and more fuel-efficient) vehicles may be less safe for occupants, and lead to increased road deaths. A National Research Council report from 2002 [11.3] found that the CAFE standards from the 1970–1980s "probably resulted in an additional 1300 to 2600 traffic fatalities in 1993." So, for a long time, there has been this tension between safety and fuel economy (carbon dioxide emissions), potentially adding to the reasons why the footprint approach was adopted in CAFE. While this tension still exists, it is perhaps diminishing as the downsides of extremely heavy vehicles become clearer.

The US is not the only jurisdiction with these incentives in its legislation. In the EU there is a 95 g/km target, measured on the NEDC, for carbon dioxide emissions for the whole EU new car fleet. This will move to 93.6 g/km in 2025, measured on the WLTP (15% reduction compared to the 2021 baseline), 49.5 g/km in 2030 (55% reduction), and 0 g/km in 2035 (100% reduction) [11.18]. However, you will note that this regulation applies to the whole EU fleet, not to individual manufacturers. Each manufacturer is given a specific target to enable the EU fleet to meet its overall objective. These targets are based on the average mass of the manufacturer's (or pool of manufacturers—many companies choose to be part of a consortium of manufacturers for this purpose) new vehicle fleet in a given year. As the EU Environment Agency itself says, "This means that manufacturers of heavier cars have higher emissions targets than manufacturers of lighter cars" [11.18]—here higher means a higher carbon dioxide emission level is permitted—that is, an easier target. This disparity can most clearly be seen in Figure 11.4, which shows the 2021 targets and achieved average emissions for the various consortia of manufacturers in the EU. Most importantly, it should be noted that every manufacturer comfortably met its target. However, the Tesla/Honda/Jaguar Land Rover consortium is notable for two reasons. First, it had by far the lowest total carbon dioxide emissions of any consortium, because all Tesla's vehicles were defined as 0 g/km by this legislation. Second, they had the highest target, driven, no doubt, because Tesla (making only luxury BEVs) and Jaguar Land Rover (making only luxury vehicles—both battery electric and ICE) have relatively heavy vehicle offerings. You can also see this with the high targets for BMW and Mercedes-Benz—both also luxury manufacturers.

FIGURE 11.4   EU average carbon dioxide emissions of pools of car manufacturers in 2021 (data from EEA) [11.18].

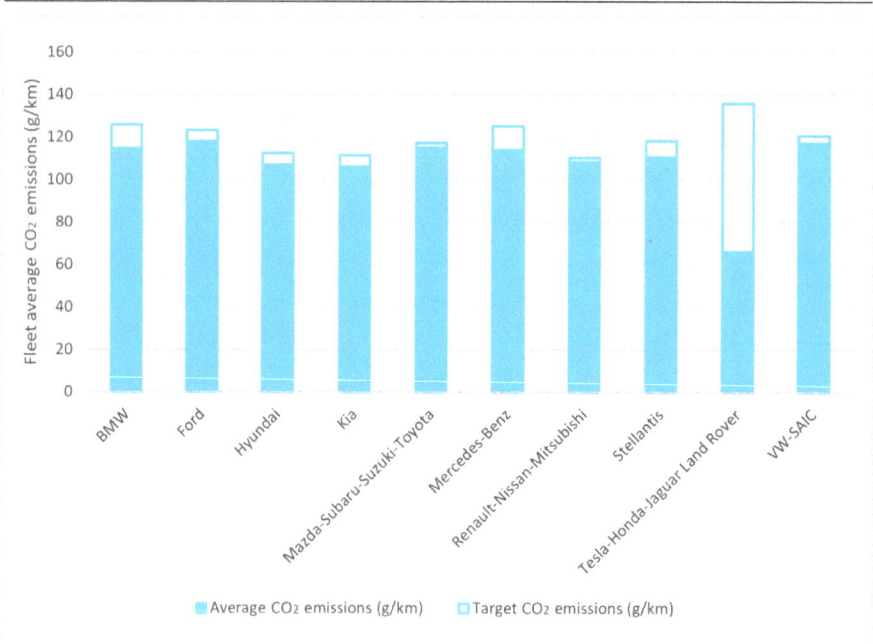

© SAE International.

It Is Not Just Legislation

We have seen that these standards permit and probably incentivize larger vehicles. Are there other reasons we are ending up with bigger cars? As regulation has increased, the amount of technology to comply with those regulations has increased and hence the cost of compliance has increased.

Larger vehicles offer more space for technological advances. This might be space for batteries in a BEV, advanced exhaust after-treatment in an ICEV, or extra sensors to provide more safety features or simply a crumple zone. Some of these are essential, rather than optional extras, for complying with regulations—and this will become increasingly so as regulations inexorably tighten.

This leads manufacturers to favor larger models for easier compliance—sometimes it may even be essential for compliance.

Compliance with the multitude of regulations often involves significant investment in research, development, and engineering. Investments by automotive companies need to be recouped. The luxury car market is where the biggest profits are to be found, and these cars are mostly (except for high-performance sports cars) big. This is why areas that have required high investment, whether advanced driver-assistance system (ADAS) features—think lane-keeping, reversing beepers, and so on—or BEVs and everything in between, have come to luxury car segments first. This is where consumers are willing to pay extra for these features, and it is where the vehicle profit margins are large. There is, consequently, a significant incentive for car companies to maximize their sales in this space. In other words: sell more big cars!

Remember that, much as it is easy to criticize companies acting in pursuit of profit, it is the fiduciary duty of their directors in law not just to survive, but to maximize value. If legislation, consumer demand, or anything else means that companies cannot make certain types of vehicles for a reasonable profit, they will need to stop making them eventually. This may be a compounding reason we are seeing this trend toward bigger more luxurious vehicles, equipped with these latest technologies on board.

Consumers Matter

We cannot, however, put all the blame for our ever-increasing vehicle sizes on legislation, standards, and the profit motive of automotive companies. We, consumers, also have a big say in this, in terms of the vehicles we buy—whether new or in the second-hand market. Consumer trends are clear: we like bigger cars. There are several reasons why this is the case.

- Safety. We have seen throughout this book the impact of vehicle size on the safety of occupants. People know this and express a preference for larger vehicles as they believe, correctly, that they provide better protection in accidents.

We have seen that this is not the whole story as vehicles with a high center of mass such as SUVs and pickup trucks may be more likely to crash, especially to roll over, in the first place. Nevertheless, the perception is still there.

- Driving position. Many larger vehicles offer a higher, commanding driving position, giving a better view of the road and a feeling of being more in control and safer.

- Status. There is a perception in many cultures that larger cars, especially large, luxury ones, project greater social status. When was the last time you saw a celebrity turn up at a red-carpet event in a compact city car? People are drawn to bigger vehicles as a symbol of success or prestige [11.19]. Part of this is brand, of course, but also a larger car is usually a more expensive car—a form of ostentatious consumption.

- Space and comfort. Bigger cars usually, but curiously not always, offer more room, providing greater comfort and space for passengers, and more storage capacity. That said, many SUVs have surprisingly small interiors given their large exteriors. Greater space is appealing for families, regular travelers, or simply people who need to move a lot of stuff. This is especially true when transporting children, with the buggy/stroller, child seat, and other paraphernalia. Furthermore, people buy a vehicle for flexibility and freedom; having a car means you do not need to rely on anyone else to travel and do the things you want to do. So, even if people do not use these amenities, the knowledge that they *could* is an important reason for owning a larger car.

- Towing and hauling capacity. Larger vehicles such as pickup trucks or SUVs typically have greater hauling capacities, making them suitable for towing trailers, caravans, boats, or other heavy loads, whether this is a reality or just comforting for the owner to know that they could.

- Adaptability to terrain. In regions with challenging terrain or weather conditions such as snow, mud, floods, or rough roads, larger cars, particularly four-wheel-drive pickups and off-roaders, are favored because of their ability to handle these conditions more effectively than smaller vehicles. This does not explain the preponderance of "Chelsea tractors" in flat, tarmacked cities, but there are clearly places in the world where this matters.

- Range. In a BEV, the range of the car is determined in large part by its battery capacity. Batteries, no matter the technology, have a relatively constant energy density (how much energy they can contain for a given size or weight). Therefore, if you want your car to have a longer range, it needs more batteries, and therefore, it needs to be a bigger car. People want long-range cars, to overcome the range anxiety often cited as a reason for being reluctant to buy a BEV [11.20]. While we can quote statistics about short average journey distances and vehicles hardly ever being driven to their potential range, people *do* value the freedom, or perception of freedom, that having that long range gives them. If you really need to, you can jump in your car, drive 400 miles, and be at the bedside of a poorly relative within a few hours, even at 3 am. That is almost impossible to achieve by any other transport type.

- Marketing. As we have seen above, car manufacturers have a big financial incentive to sell larger vehicles, so these are the ones that they market. In addition to the financial motive, it makes sense that car companies will often market larger vehicles with attractive features, amenities, and designs that appeal to consumers' preferences, further increasing the desire for bigger cars. When was the last time you saw an advertisement for the cheapest or smallest vehicle in a manufacturer's lineup?

A final consideration when buying a car is its resale value. Even if an individual is not driven themselves toward a larger vehicle by the above considerations, they may still err that way if they think other people prefer them. People want to play safe and buy a car that they think will hold its value and for which there will be lots of potential buyers—in economic terms, they wish to retain the greatest liquidity for their asset. Therefore, there will be a tendency to "follow the crowd." This is a reason why so many cars today are of similar colors. You *could* buy one new in bright pink, but such a vehicle is likely to be less desirable on the second-hand market. This is a second-order point, but may influence purchasing decisions.

The Results of All This

Where, then, does this leave the state of the car market today? As you will have guessed, there are more and more bigger cars on the roads and being sold today than ever before, and the trend is accelerating. Figure 11.5 shows the market share of different types of vehicles in the US from 1975 to the present [11.21]. It is clear that larger vehicles, particularly light trucks, are now dominant and that vehicles that would typically be considered a "car" now make up less than 30% of sales. Big vehicles are already here, and we are all living with the consequences.

FIGURE 11.5   Market share of different vehicle types in the US, 1975–present (data, US EPA) [11.21].

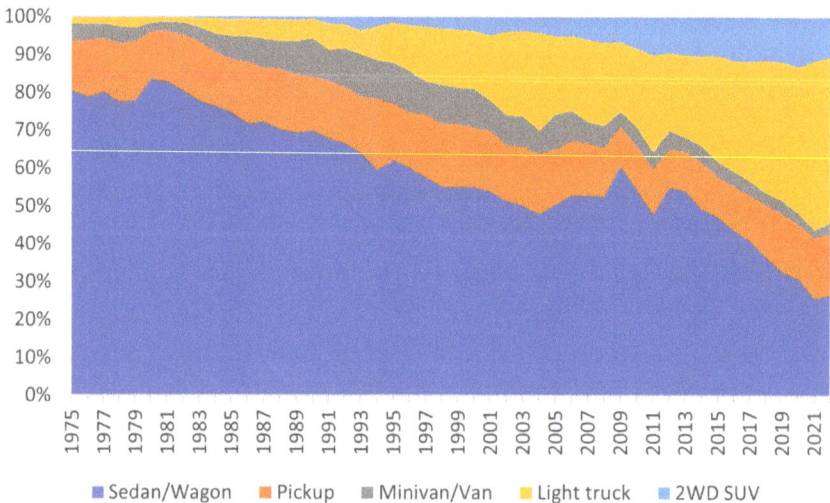

Sedan/Wagon   Pickup   Minivan/Van   Light truck   2WD SUV

© SAE International.

It is logically true that a bigger car is not *necessarily* a heavier one. However, it is in reality. Figure 11.6 shows the average vehicle mass of cars in the US over the same period [11.21]. The effects of the oil crisis inducing lightweighting are clear in the late 1970s, but since then the trend has been a steady move upward. The average car today weighs over half a tonne more than the average car in 1980, and there has been a 20% increase in the last decade. This is *before* the impact of BEVs is really felt in the US market (1.2% of new vehicle registrations in 2022 [11.22]), which will only increase the problem further, and substantially so [11.23].

FIGURE 11.6 Average vehicle mass in the US market 1975–present (data, US EPA) [11.21].

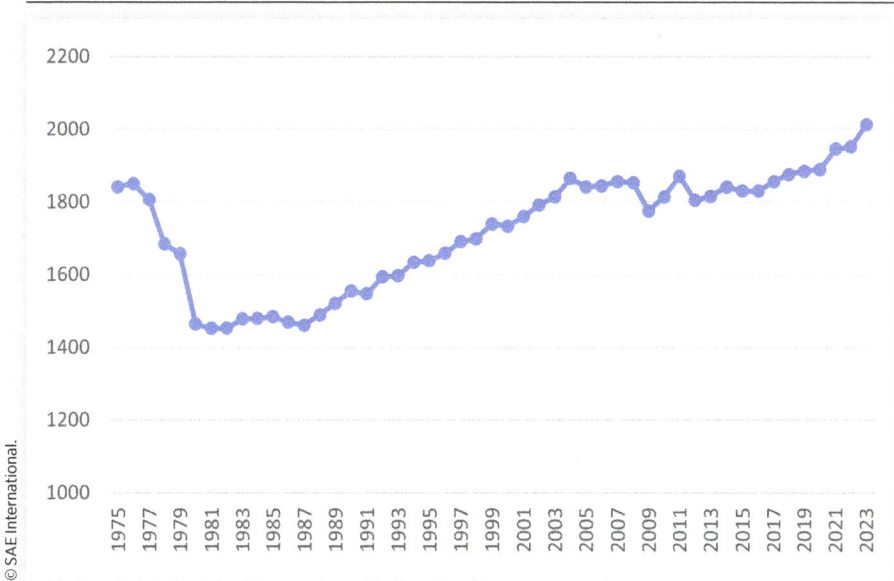

© SAE International.

If you do not live in the US, then you might think, stereo-typically, and perhaps a little prejudicially, "Ah well, everything is bigger over there." However, the trends are the same world-wide [11.24]. Yes, the US has the heaviest cars worldwide—see Figure 11.7. Among developed economies, there is a big range— France has an average weight 25% lower than the US, and Canada and Japan have an average vehicle weight 33% lower. Nevertheless, over time, we are all catching up with the US and Canada—see Figure 11.8, which shows the weight gain of vehicles between 2011 and 2020. In particular, if you look at Figure 11.8, the trend of getting heavier is abundantly clear—with increases in mass notable in Austria and Sweden of the countries shown. Many countries, including France, have seen increases greater than in the US over this period.

FIGURE 11.7 Average vehicle masses in various countries worldwide [11.24].

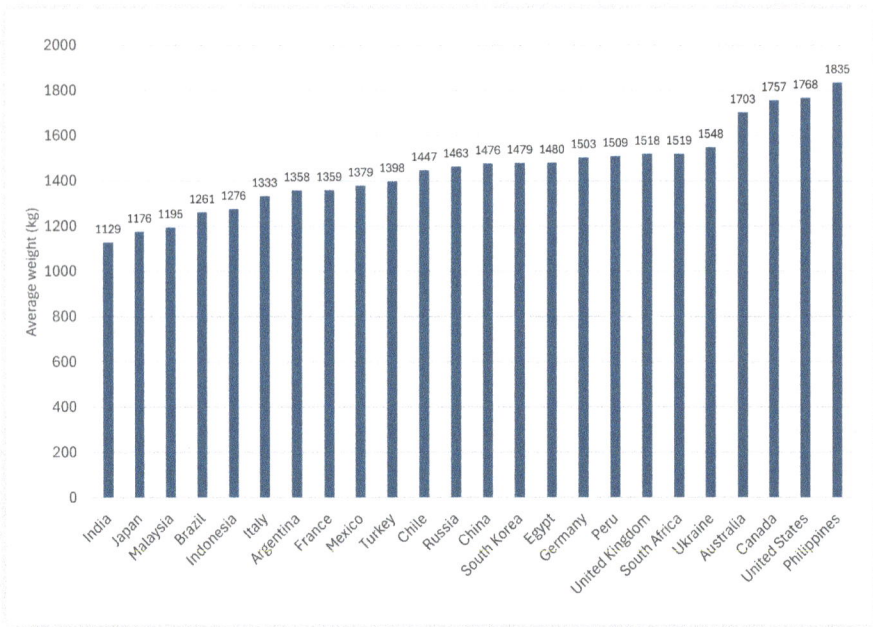

© SAE International.

FIGURE 11.8 Increases in car mass for selected countries worldwide between 2011 and 2020 [11.24].

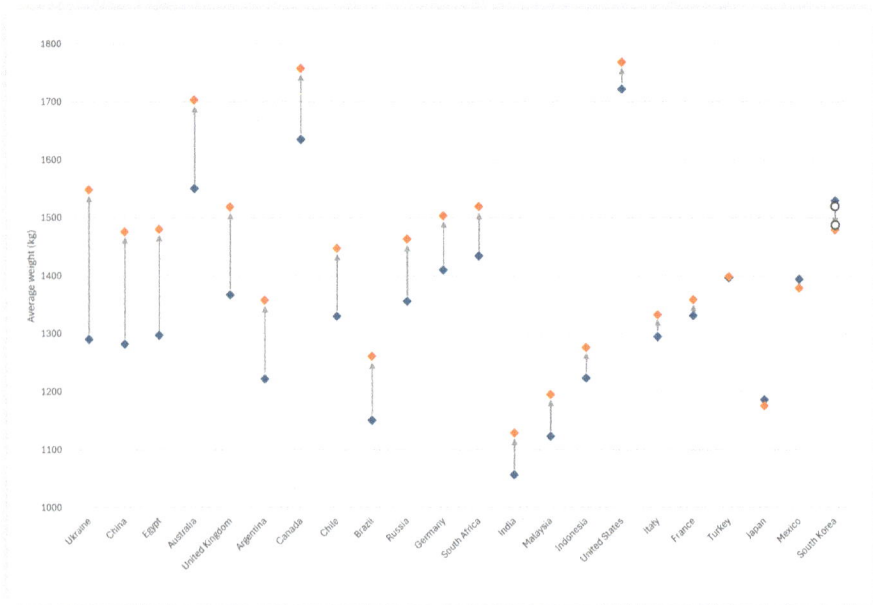

© SAE International.

Which Cars Are Putting on Weight the Most?

Throughout this chapter, we have been placing particular emphasis on, and implicitly critical of, the move to larger cars, but the interaction of causes is not clear. To what extent is it a shift of consumer preference, or is it a product push from manufacturers? Is it because people are buying bigger cars or is it that when an individual model of a car is updated, it gets bigger? We have already seen that the trend toward buying more light trucks relative to cars is part of the issue (Figure 11.5), but that is not the only reason. Individual models of cars are also getting bigger over time. We already saw back in Chapter 1 that the Volkswagen Golf has roughly doubled in weight over the 50 years it has been in production. That is not unique. Figure 11.9 shows the sizes of the early models of the Mini, Porsche 911, and Fiat 500 alongside their current production models. The increase in size is abundantly clear. As we have already noted, the increase in size leads to an increase in weight. The original Mini Mk 1 Saloon weighed 587 kg, the 2024 model's three-door hatch weighs 1260 to 1680 kg depending on options, and the Mini Electric weighs 1775 kg—this is really no longer living up to its name! The original Porsche 911s weighed 900 to 1100 kg (depending on exact specifications), and the 2024 model weighs 1455 to 1785 kg; the original Fiat 500 weighed 499 kg, the 2023 ICE model weighs 865 to 1149 kg, and the 2024 battery electric model (Fiat 500e) weighs 1255 to 1405 kg (the ICE model is no longer on sale).

FIGURE 11.9 Increases in size between older and current models of Mini (top), Porsche 911 (middle), and Fiat 500 (bottom).

Giovanni Love/TamasV/Shutterstock.com.

Courtesy of Paul Steinbruner.

Giovanni Love/TamasV/Shutterstock.com.

These are not isolated examples. We have taken the smallest vehicles offered by the world's top five car manufacturers (by volume) in 2024 who were (in descending order) Toyota (including Lexus), Volkswagen (who also make brands including Audi, Bentley, Bugatti, Lamborghini, Porsche, SEAT, and Škoda), Hyundai-Kia, Stellantis (including Alfa Romeo, Chrysler, Citroën, Dodge, Fiat, Jeep, Lancia, Maserati, Opel, Peugeot, Ram, and Vauxhall), and General Motors (including Chevrolet, GMC, Cadillac, and Buick). The weights of these cars are shown in Figure 11.10. Except Volkswagen, whose smallest car has lost 6.5% of its weight, every other smallest car has got heavier, with the average weight gain across all brands (including Volkswagen) being just over 10%. What is also notable is that, for Stellantis, its lightest vehicle has been discontinued and the Citroen C3, a B-Segment vehicle, is now its lightest vehicle. Its A-segment vehicles are now BEVs and heavier, as we have just seen with the Fiat 500e.

FIGURE 11.10   Weights of the lightest cars available from the world's top five manufacturers (by volume) 2010 (blue) and 2024 (red).

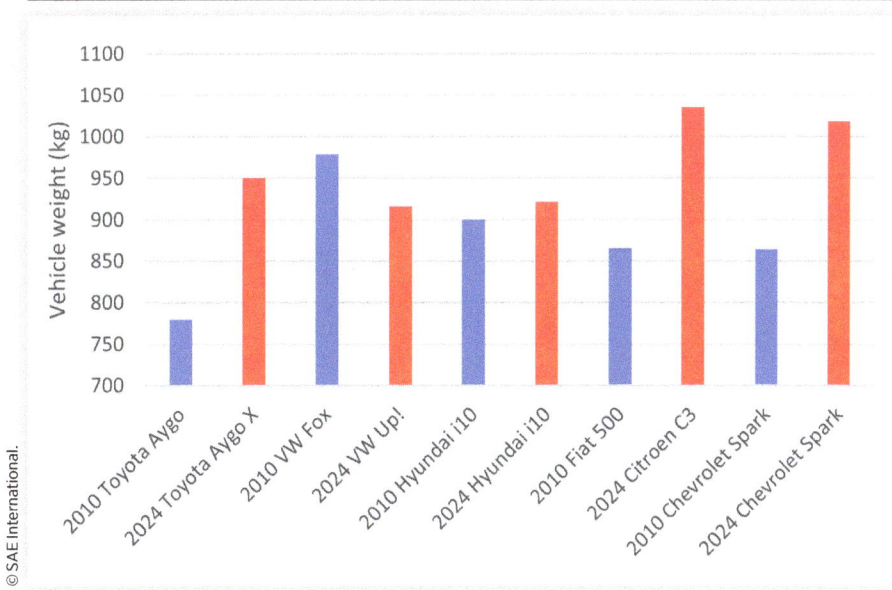

© SAE International.

We have also taken the heaviest vehicles sold by each brand over the same time period, and added for 2024 both the heaviest ICE-powered model and heaviest battery electric model—see Figure 11.11. Again, there is substantial variation between manufacturers, but, except Stellantis and its enormous RAM 3500 pickup truck that has lost 5% of its weight, the heaviest vehicles have all got heavier over the same period. The average weight gain for the heaviest ICE-powered model is 8%. For battery electric models, the picture is more mixed, with some manufacturers not yet having released a battery electric equivalent for their heaviest model, and the borderline insanity of the weight of the Hummer EV skewing the numbers. Nevertheless, compared to 2010, when there were no battery electric models on sale, the average weight gain across the five manufacturers is 13%.

FIGURE 11.11   Weights of the heaviest ICE cars available from the world's top five manufacturers (by volume) 2010 (blue) and 2024 (red) and 2024 heaviest battery electric models (green).

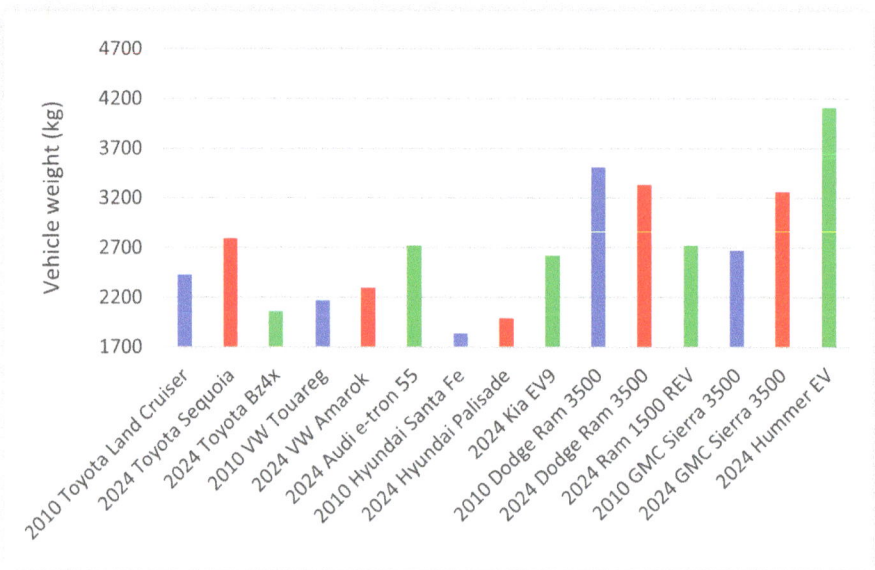

© SAE International.

Even for cars which are usually designed to be as light as possible—sports cars—we see some of the biggest proportionate weight gains. Figure 11.12 shows the weights of selected sports cars in 2000 (or 2003 for Mercedes-Benz as we could not find a direct 2000 equivalent for this model) and today's equivalents. Here, the trends are clearest—not a single sports car has got lighter. Indeed, on average, there is a 48% rise in weight. Some of the comparisons are startling, for example with the Lotuses having more than doubled in weight. No doubt there will be particularly passionate objections to some of the precise comparisons here, not all sports cars are equal of course, but the pattern is unmistakable.

FIGURE 11.12   Selected sports car weights 2000 (blue) and 2024 (red) equivalent.

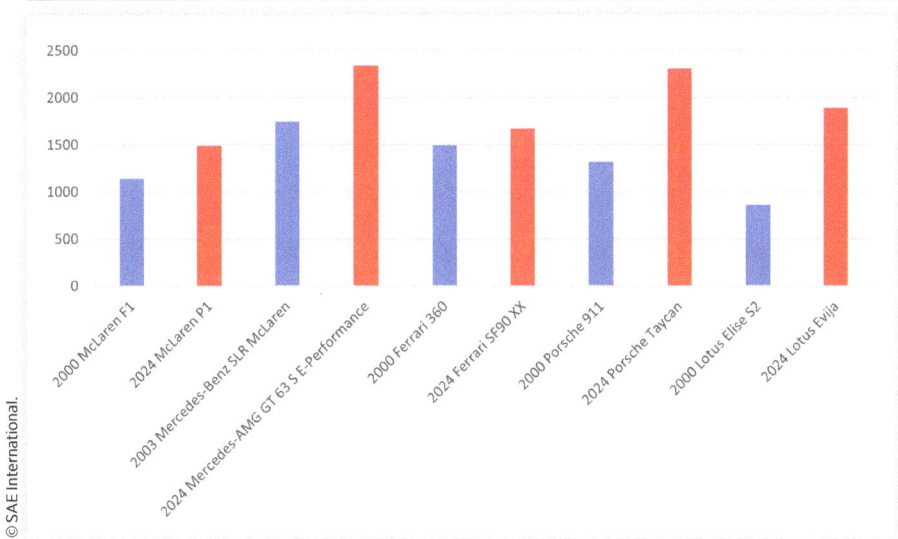

© SAE International.

Summary

Overall, then, the trends are clear. Our average car is much heavier than it used to be, and it is not just because consumers are buying bigger cars. Equivalent models are also getting bigger and heavier. With a few notable exceptions, across every segment from smallest to largest, including sports cars—where lightweight is a key selling point—equivalent models are getting bigger and heavier. The automotive obesity pandemic is here.

12

The Proposal

IN THIS CHAPTER

- We propose a tax based on unladen vehicle mass multiplied by distance driven.

- How this would work in the UK, California, Germany, France, and Japan is demonstrated.

- We show why this approach is superior to alternative taxation structures.

So, we reach the moment you have all been waiting for: how we can go forward and efficiently label and tax vehicles in this new diverse powertrain world?

Why do we *need* to do this? Our proposal here is not a fanciful "nice-to-have." As we remove ICEVs from the market, sales of gasoline and diesel will go down. Demand has already peaked in Europe and the US and is expected to peak globally between 2027 and 2030—see the forecasts in **Figure 12.1** [12.1, 12.2]. Why does this matter? Almost every country worldwide taxes fuel for road vehicles in some way. Such taxes have, historically, as we have seen, been a good surrogate for the environmental impact of the use of vehicles, including the pollution from fossil fuel production. While you can argue that fuel taxes are not high enough (if you are concerned about the pollution from vehicles and fossil fuel use) or are too high (if you are concerned about the cost of mobility and the regressive nature of fuel taxes), we are not here to have that fight. However, it is unlikely that governments, many of which are now giving subsidies to promote uptake of BEVs, are going to give up happily the revenue from fuel taxes. While fuel taxes are currently a good surrogate for the environmental impact of vehicles, this will change with the evolution of the powertrain mix, so we need to update the taxation regime accordingly.

FIGURE 12.1 Predictions of global road fuel demands worldwide from BNEF and Goldman Sachs [12.1, 12.2].

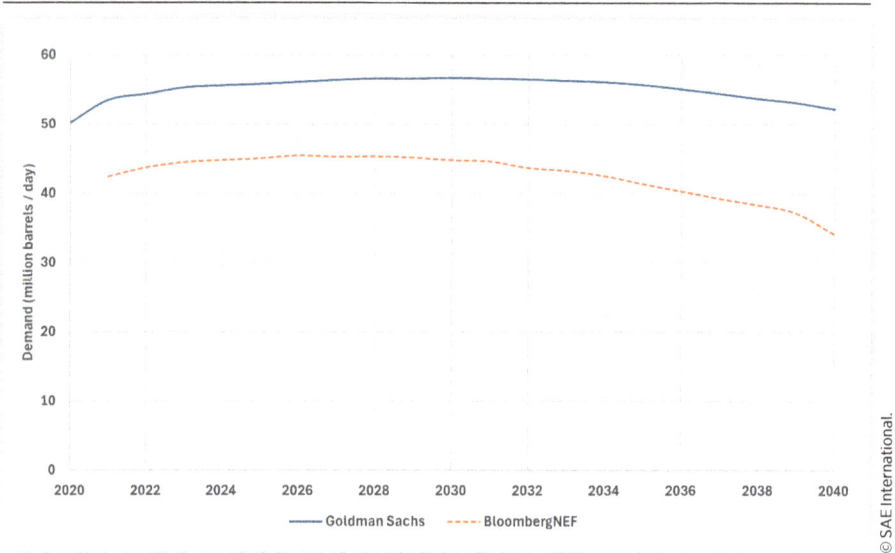

The current tangle of global vehicle taxation is covered in Chapter 4. Despite being a mess, it serves as the benchmark for determining what revenue governments will want or need to generate. The simplicity of what we will propose means that you can apply our method to different countries with different political and taxation philosophies. For example, some countries, particularly in Europe, tax fuel and vehicles at relatively high levels (in common with taxes on many other things), whereas others, such as the US, have lower taxation levels. The proposal can be tailored to individual countries and their political and taxation outlooks.

Our proposal is not a complete solution for every country, the details of which are the responsibility of national governments and any devolved administrations as appropriate. Rather, we propose a framework for moving from a patchwork of car and fuel taxes to a new simpler system. We will illustrate this with reference to selected countries where we propose a revenue-neutral transition from one system to another, using the publicly available information.

It is worth reminding ourselves about current taxation systems and what they actually tax. Most have some sort of annual tax on the vehicle itself, plus a tax on fuel. The tax on the vehicle itself is a flat rate in some jurisdictions and variable in others. Variable taxations come in a wide range of types; these include working on the basis of engine power, a percentage of vehicle value, the vehicle's carbon dioxide emissions, vehicle age, or engine size, although there are many other combinations. Tax on fuel is effectively a levy per mile driven. As burning fuel produces the emissions we are concerned about, and these emissions are broadly in proportion to the amount of fuel burnt, the tax also acts as a pollution levy. A further property of this tax is that it acts as a penalty for less efficient vehicles, which will often be heavier vehicles. So, now, most countries in some way or other tax negative characteristics of vehicles with a form of activity tax [12.3].

The *Molden-Leach Conjecture*

Our proposal takes the existing situation and updates it for a new diverse powertrain world, as set out so far in this book and the *Molden-Leach Conjecture*. To this end, **we propose a tax based on the product (multiple) of unladen vehicle mass and distance driven**. In other words, how many kg × km or lb × miles you do in a given year. The heavier your car and the more you drive, the more you pay.

Why is this proposal intuitive? Well, the scientists—professional or otherwise—among you will note that the units of this product are the same as for moments or torque. On the surface of it, this may sound a bit weird. However, when you consider that (on Earth and all planets with nonnegligible gravity) mass is, in reality, a force, or weight, then the units become force × distance. And force × distance is also work done or energy. Therefore, we are taxing energy expended.

As we have seen throughout this book, energy is responsible for the environmental impact of vehicles, regardless of their type, in many ways, whether that is their life cycle production or disposal environmental impact, their in-use environmental impact (where the energy used to move them matters), their safety impact on others, and so on. A full list of reasons have been discussed elsewhere in this book. Suffice it to say, energy is a sensible unit for us to be considering taxing vehicles on.

Rather than talking in the abstract, we think it is easier to explain with an example. We are both British, so we will start with a UK example.

UK

The UK taxes vehicles in a number of ways, including import duties on foreign-manufactured vehicles, VAT on purchase, and tax write-downs for depreciation. We do not need to touch these or suggest that they would change as the powertrain mix changes in the future, so we will not. That helps keep things simple, which is key. The three relevant taxes on light-duty vehicles are as follows:

1. VED—an annual tax that is paid for owning a car and being permitted to drive it on the road [12.4].
2. Fuel duty—a tax per liter of fuel sold [12.5].
3. VAT on fuel duty—yes, a tax on a tax [12.5].

Vehicle excise duty until 2001 was a tax based on engine size. From 2001 to 2017, it was based on the carbon dioxide emissions of the vehicle. Since 2017, it has been based on the carbon dioxide emissions of the vehicle in the first year of its registration and a flat fee after that. In addition, there is a top-up tax for vehicles with a purchase price over a certain level. BEVs do not pay this tax whatever their price, although this will change in 2025 when the standard rates, including the top-up for higher-priced vehicles, will apply [12.4].

Fuel duty in the UK is 52.95 pence/L for both gasoline and diesel, a rate which has been constant in nominal terms since 2011. VAT is applied on top of this, which is currently 20%. So, the total duty including VAT is 63.54 pence/L [12.5].

In common with most other products in the UK, the 20% VAT is charged on the net sale price of the gasoline or diesel and therefore varies with that net price. VAT is also charged at the same rate on the sale of electricity per kWh at public charging points. As it is unlikely that a government will want to increase or decrease the rates of VAT on either gasoline, diesel, or electricity, or indeed change the VAT system itself beyond tweaks, we have assumed these are constant in our calculations. As a side note, if you charge up your car at home, the VAT rate on domestic electricity is just 5%; as evidence of the challenges of changing VAT rates, the UK government has not (yet) aligned this domestic rate with the public charging rate. Given that the average cost to charge an electric vehicle is lower than the cost to get an equivalent range from gasoline or diesel once you have stripped out the fuel duty, it is likely that significant revenue will be lost in the transition from liquid fuels. We have not accounted for this.

So, in the UK, the tax revenue that will be lost by a complete transition to electric cars will be mostly from vehicle excise duty, fuel duty, and VAT on the fuel duty. How much revenue is this? We considered the tax year 2019–2020 because, at the time of writing, data for post-pandemic years is not yet fully available. It is worth bearing in mind that the effects of high inflation since then will have eroded the real value of fuel duty, as it is expressed in monetary terms and has not been increased in line with inflation, such that 2024 revenues will be somewhat lower in real terms. The total revenues taken from the selected year are shown in Table 12.1—all data is taken from the UK's Office for Budget Responsibility (OBR) [12.6, 12.7]. The total, £39.9 billion, made up 6.3% of all tax raised in 2019–2020, which makes it a significant portion of the total government income in the UK.

TABLE 12.1 Tax revenues raised from selected vehicle taxes in the UK, 2019–2020.

Tax	Revenue (2019–2020)
Fuel duty [12.7]	£27.6 bn
Vehicle excise duty [12.6]	£6.8 bn
VAT on fuel duty at 20%	£5.5 bn
Total	£39.9 bn

If the intention of our proposal is to match the revenue, to avoid questions of whether tax overall should be higher or lower, we need our new system to raise £39.9 billion.

At first glance, then, you might consider the following approach. We know the number of vehicles in the UK (approximately 41.3 million) [12.8], the average annual miles per car (approximately 6600 miles) [12.8], and the average unladen mass of a vehicle in the UK (1500 kg) [12.9]. With that, we can work out that we would need to charge approximately 9.7 pence per tonne per mile driven (or £9.71 $\times$ 10^{-5} per kg $\times$ mile). We can obtain the unladen vehicle mass (kerb weight) from when the vehicle is first registered (for UK readers, you will find it on your V5C vehicle registration form—it is listed as "revenue weight") and the mileage from the annual inspection of the vehicle (the so-called MOT test), which is recorded and logged automatically with the Driver and Vehicle Licensing Agency (DVLA, the UK licensing authority for vehicles). This is also recorded whenever a vehicle is bought or sold. Eagle-eyed UK readers will note that a car only needs its first MOT test three years after the first registration date, so we will need another way of recording the mileage in those years. A voluntary system during this period, which is then cross-checked at the three-year mark or at sale, would probably work.

We have done it! A new system. About 10 pence per tonne per mile.

But there is a technical problem. Suppose you converted our units to kg $\times$ km or lb $\times$ mile, as would be common in Europe and the US, respectively). For each of these, you will get a different answer. Why? Because what we have inadvertently done is apply a weighting to distance and mass. If you consider the average vehicle, which would have a mass of 1500 kg and travel 6600 miles, we have implicitly weighted the mass factor by 19% and the distance factor by 81%. If you did it in lb $\times$ miles or kg $\times$ km, the mass-to-distance weights are 33%/67% and 12%/88%, respectively. We cannot have a system where the answer changes depending on what units you may select, not least because the system would not apply universally across countries using different units.

You also may wish to choose a weighting yourselves for the relative contribution of mass and distance. To do that rigorously, you would need to know the relative contribution of mass and distance to the environmental impact of vehicles, and that is going to be both hard to calculate and ultimately subjective. It is of interest to note that the system in the UK today, assuming you drive an average distance annually in an average car, effectively weights distance at approximately 65% and simply owning a car at 35%.

So, we propose a simple 50/50 weighting. Yes, it is a choice, but it seems the fairest way to juggle these two contributions. This system gives motorists a balanced choice: the same environmental benefit can be achieved either by reducing vehicle weight by 10% or driving 10% less distance. How does this work in practice? Well, you now need to scale the vehicle mass and the annual distance driven relative to the averages for your country (or some other target values if your goal is to bring either average vehicle mass or distance driven down). This is how it would work:

$$\text{Annual vehicle tax} \, (\pounds) = \frac{m}{m_{av}} \times \frac{d}{d_{av}} \times \frac{T}{N}$$

where, in our UK example, m is your vehicle mass (in kg), m_{av} is the average mass (in kg) of vehicles in your jurisdiction (or the target mass that you would like to achieve), d is your annual distance driven (in miles), d_{av} is the average distance driven (in miles) of vehicles in your jurisdiction (or the target annual mileage that you would like to achieve), T is the tax that you would like to raise (in £), and N is the number of vehicles that pay that tax each year.

This formula is applicable worldwide in any jurisdiction, just the units, averages, number of vehicles, and tax you wish to raise would change. So, what would this UK example look like in reality? A few example vehicles of different types and mileages are shown in Table 12.2 for illustration using estimated inputs.

You can see from Table 12.2 that there are winners and losers, the heavy vehicles that drive a long way—the Jeep Grand Cherokee, for example—are heavily penalized, but lightweight cars that do not go far in a year end up being taxed lower. Those vehicles are likely, as we have shown in previous chapters, to have a lower environmental impact, so our system incentivizes these even further. Note that this is only a selection of vehicles, but across the market the system is revenue neutral.

TABLE 12.2 Example taxations for popular vehicles in the UK market under this new mass–distance-based system and a representative current figure shown.

Vehicle	Weight (kg)	Mileage (indicative)	Tax under existing system	Tax under new system	Increase in tax
VW Up!	980	5000	£456	£490	£34
Mazda MX-5	1066	3000	£456	£320	-£136
Ford Puma	1358	4000	£866	£543	-£323
McLaren 720S	1419	2000	£2715	£284	-£2432
Renault Zoe	1502	5000	£0	£751	£751
Škoda Yeti	1565	9000	£737	£1409	£672
Nissan Qashqai	1670	8000	£1087	£1336	£249
Jaguar XF	1800	12,000	£1233	£2160	£928
Tesla Model Y	1930	10,000	£0	£1930	£1930
BYD Seal AWD	2185	6000	£0	£1311	£1311
Jeep Grand Cherokee	2503	10,000	£1112	£2503	£1391
Kia EV9	2660	8000	£0	£2128	£2128
Audi SQ8	2725	4000	£2826	£1090	-£1736

This approach can be applied worldwide, provided you have the four key bits of information needed:

1. How much tax revenue you would like to raise.
2. The number of vehicles that will be subject to this tax.
3. The average (or target) distance driven each year by those vehicles.
4. The average (or target) mass of those vehicles.

We have obtained estimates of these for a number of other jurisdictions worldwide and have sought out the most popular vehicles there to give relevant examples. This is a generic system by design, so it can work anywhere!

California

For the California, US, example, the values we have put into the calculation are shown in Table 12.3. These are illustrative of the Californian market. The resulting tax values for ten popular vehicles in this market are shown in Table 12.4.

TABLE 12.3 Values required for the calculation of the new tax system for California, US.

Average vehicle mass	4000 lb [12.9]
Average annual distance driven	12,524 miles [12.10]
Total number of vehicles taxed	14.3 m [12.11]
Tax revenue raised	$20.9 bn [12.12]

© SAE International.

TABLE 12.4 Example taxations for the top-selling vehicles in the California, US, market under this new mass–distance-based system.

Vehicle	Weight (lb)	Mileage	Tax under new system
Toyota Camry	3100	14,000	$1269
Honda Civic	2877	6000	$505
Toyota Prius	2903	22,000	$1867
Tesla Model Y	4416	20,000	$2582
Toyota Corolla	3055	14,000	$1251
Nissan Altima	3373	8000	$789
Toyota RAV4	3655	10,000	$1069
Ford Focus	2640	7000	$540
VW Jetta	3036	19,000	$1687
Tesla Model 3	4034	6000	$708

© SAE International.

France

In France, the values we have put into the calculation are shown in Table 12.5. These are representative of the French market. The resulting tax values for ten popular vehicles in the French market are shown in Table 12.6.

TABLE 12.5 Values required for the calculation of the new tax system for France.

Average vehicle mass	1400 kg [12.9]
Average annual distance driven	13,117 km [12.13]
Total number of vehicles taxed	38.9 m [12.14]
Tax revenue raised	€60 bn [12.15]

© SAE International.

TABLE 12.6 Example taxations for the top-selling vehicles in the French market under this new mass–distance-based system.

Vehicle	Weight (kg)	Annual km	Tax under new system
Peugeot 208	1160	16,000	€1561
Dacia Sandero	1237	6000	€624
Renault Clio	1065	22,000	€1970
Citroen C3	958	14,000	€1128
Peugeot 2008	1300	14,000	€1530
Renault Captur	1759	8000	€1183
Dacia Duster	1250	10,000	€1051
Toyota Yaris	950	10,000	€799
Renault Twingo ZE	1112	18,000	€1683
Ford Puma	1760	14,000	€2072

© SAE International.

Germany

The values for Germany we have put into the calculation are shown in Table 12.7. These are representative of the German market. The resulting tax values for ten popular vehicles in the German market are shown in Table 12.8.

TABLE 12.7 Values required for the calculation of the new tax system for Germany.

Average vehicle mass	1577 kg [12.16]
Average annual distance driven	13,602 km [12.17]
Total number of vehicles taxed	48.8 m [12.14]
Tax revenue raised	€84 bn [12.15]

TABLE 12.8 Example taxations for the top-selling vehicles in the German market under this new mass–distance-based system.

Vehicle	Weight (kg)	Annual km	Tax under new system
VW Golf	1465	16,000	€1882
VW Tiguan	1604	6000	€773
VW T-ROC	1555	24,000	€2997
Fiat 500	1365	17,000	€1864
Opel Corsa	1246	12,000	€1201
Mini	1610	8000	€1034
Ford Kuga	1716	7000	€965
BMW 3 series	1590	10,000	€1277
Tesla Model Y	1998	20,000	€3209
Mercedes C-class	1615	16,000	€2075

Japan

Japan is famous for its small vehicles with their light weights—the kei cars—so its average vehicle mass is very low. The values we have put into the calculation for the Japanese market are shown in Table 12.9. The resulting tax values for ten popular vehicles in the Japanese market are shown in Table 12.10.

TABLE 12.9 Values required for the calculation of the new tax system for Japan.

Average vehicle mass	1176 kg [12.9]
Average annual distance driven	6790 km [12.18]
Total number of vehicles taxed	62 m [12.19]
Tax revenue raised	¥8.7 tn [12.20]

TABLE 12.10 Example taxations for the top-selling vehicles in the Japanese market under this new mass–distance-based system.

Vehicle	Weight (kg)	Annual km	Tax under new system
Toyota Yaris	1110	9000	¥175,698
Toyota Sienta	1270	3000	¥67,008
Nissan Note	1190	12,000	¥251,148
Toyota Roomy	1405	10,000	¥247,103
Toyota Prius	1317	7000	¥162,138
Honda Freed	1380	4000	¥97,082
Toyota Harrier	1700	5000	¥149,493
Toyota Land Cruiser	2500	5000	¥219,842
Toyota Noah	1650	10,000	¥290,192
Toyota Voxy	1600	8000	¥225,118
Honda N-Box	920	6000	¥97,082
Toyota Corolla	1100	4000	¥77,384
Honda Fit	1030	2000	¥36,230

© SAE International.

All these examples are based on assumptions, but these are set out and you are free to play around with them. What we have set out to illustrate is the framework and how it can be applied to any context. We have also made a simple spreadsheet tool, available online, http://dx.doi.org/10.5287/ora-zb9yndz9n, to enable you to do these calculations for yourself anywhere. This method can be applied in any jurisdiction from Afghanistan to Zimbabwe provided you have the key bits of information listed above.

Effects

What, then, would be the effect of this new tax? Given the substantially increased penalization of heavy cars, we would assume that this would, in time, incentivize smaller cars, leading to a shift in consumer demand away from heavy cars. This should reverse the trend of recent decades, as we have seen in this book. In the short term, in a market where the existing vehicle fleet is dominated by ICEVs, it would reduce fuel use and emissions as well. As and when BEVs become the dominant technology, it would also reduce the electricity that they require and hence emissions associated with their generation. We would also anticipate non-exhaust and LCA emissions to reduce.

Additionally, we expect that taxation in this manner would further incentivize the development of more energy-dense batteries appropriate for the automotive sector, as well as other technologies such as lightweight new materials, for example. Although we note that the incentives to develop more energy-dense batteries are already high, there is a huge demand for these in many sectors where battery weight is vital, such as in aviation.

Aside from incentives for mass reduction, there is of course also the incentive created by this new tax system simply to drive fewer miles. With traditionally-fueled vehicles, the incentive is already there because of fuel tax and VAT, but with electric vehicles and their lower cost of electricity—especially when charged at home—the incentive to drive less is much weaker, even though they contribute to pollution and congestion at least as much. Our system at the very least reasserts this incentive to drive fewer miles and possibly increases it depending on the current taxation system you live in.

Alternatives

It is worth considering alternatives to our proposal. By alternatives, we mean maintaining the tax base from vehicles rather than any other policy goal to increase or reduce that burden. In other words, how else can governments recoup the revenue likely to be lost from the disappearance of fuel taxation in the coming years? As we see it, there are two main alternatives:

1. Tax electricity and sustainable fuels that go into a vehicle at a rate equivalent to that of fuel duty.
2. Road pricing—taxing vehicles for the use of roads.

In our opinion, commenting only at a high level, neither of these alternatives will work, for two different reasons.

In the case of taxation of electricity, it is going to be very difficult to implement. It will be almost impossible to know whether a unit of electricity is being used to charge a car or to power a light bulb, charge your phone, or whatever. Even if you could meter separately, clearly it would be possible to circumvent such a system by trailing a lead out of a plug socket from your house. Your car would charge more slowly, but many would consider it worthwhile to avoid tax. If you applied a tax equivalent to fuel duty to *all* electricity, we suspect it would create a significant political backlash. The UK used roughly 275 TWh of electricity in 2022 [12.21]. To cover the £39.9 billion of lost revenue, a charge of roughly 14.5 pence/kWh would be needed. This assumes electricity use stays flat despite increased BEV charging—definitely not true, but we are in the right ballpark. This would raise the annual bill for an average household by £391.75 and roughly increase the cost of electricity per kWh by 50% at 2024 prices. Given the recent substantial inflation in energy costs, compared to 2020 prices, the cost of electricity would have approximately doubled in cash terms [12.22]. Clearly there would need to be an equivalent increase in the costs of many goods and services, as all businesses and manufacturing companies would need to be subject to the same tax on their electricity usage.

This, therefore, is a nonstarter.

Road pricing, also known as road user pricing (RUP), congestion pricing, or tolling, involves charging drivers for the use of roads based on factors such as time of day, congestion levels, and distance traveled. While road pricing has been implemented successfully in some places—for example the congestion charge in central London has been running successfully since 2003 [12.23] and Singapore's Electronic Road Pricing (ERP) has too since 1998 [12.24]—it faces challenges and opposition in many regions, especially when deployed at large scale. These challenges are mostly around privacy concerns and political sensitivities.

The technology required for road pricing often involves tracking the movements of vehicles, raising privacy concerns among the public [12.25]. Some individuals may be uncomfortable with the idea that their location is being monitored for the purpose of tolling. Others may be fine with that, but then worry about how this information is stored and what else it may be used for. Holding tracking data on huge swathes of the population raises significant privacy issues—particularly if the information were accessed in a large-scale hack. Similarly, would the police be able to prosecute on the basis of it? What about under more authoritarian governments? It would be easy to block access completely to a certain part of the country and track anyone who broke that rule. For these reasons alone, it is our opinion that comprehensive road pricing will probably never be implemented.

Road pricing can also be politically sensitive, as elected representatives may fear backlash from the public [12.26]. People are likely to get annoyed when road pricing is implemented as they are going to have to pay for something that they have never had to pay for before. It is also likely that road pricing would be implemented in such a way that it incentivized traveling at less popular times to manage congestion. Imagine the disquiet when popular times—traveling in the holiday period, driving during the school drop-off times—became much more expensive, especially for those with little flexibility and on lower incomes. You may think that such an approach is justified or even desirable, for example in terms of economic efficiency, but you can also understand why elected officials may be hesitant to support such measures.

A third, but usually less problematic reason, why we do not think that road pricing will work is around the cost of implementation. For example, the Ultra Low Emissions Zone expansion in London (which charges some vehicles based on their Euro emissions standard) cost £130–140 million to implement [12.27] although the charge is set high enough that it is expected to generate £200 million in net revenue in its first two years of operation [12.28]. Needless to say, it has been immensely politically controversial [12.29, 12.30]. Implementing road pricing often requires advanced technological infrastructure, such as electronic tolling systems or GPS tracking. These systems can be costly to implement and add on administration and enforcement costs, and the tax take required simply to cover those costs can be significant. Given a lack of public support, policymakers may be reluctant to invest such sums up-front.

A clever alternative proposal was made in a report from the Policy Exchange think-tank in the UK, and which won the Wolfson Economics Prize in 2017: *Miles Better. A distance-based charge to replace Fuel Duty and VED, collected by insurers* [12.31].

Rather than road pricing, it uses a distance metric similar to what we propose, and suggests using insurers as an efficient method of collection. The main weakness in the proposal is that the pollution externality is measured as a weighted combination of factors, and embeds arbitrary preferences for certain types of vehicles. Their proposal, combined with our mass-based emissions rating, could work effectively.

Summary

Therefore, we propose a simple, globally applicable formula to tax cars based on their environmental impact by using their mass and distance driven in a year. We have shown how this would look in a number of different global jurisdictions and offered a simple tool to enable you, the reader, to make your own scenarios. Looking at the alternatives, assuming that governments will want to retain tax revenue from cars, the alternatives that we have considered seem to be unviable for different reasons. As we move to this new, diverse powertrain world and ultimately toward full electrification, we see no sensible alternative to our proposal, which will capture simply, yet effectively, the environmental externalities of our car usage.

13

Deployment, Limits, and Secondary Effects

IN THIS CHAPTER

- We consider the limitations to the *Molden-Leach Conjecture*, and potential unintended consequences.

- Fossil fuel taxation would need to be retained for the system to work universally between combustion and electric vehicles.

- The proposal does not claim to be perfect, but it is simple and good enough.

So, we have our proposal for how best to account for the environmental impact of vehicles in a new diverse powertrain era, and we have explored how it might work using examples from three continents. But we are not done quite yet. To recap, this book has not been about creating a highly complex rating that encompasses every environmental impact—quite the opposite. As we have established in previous chapters, accounting properly for the environmental impact of all vehicles is an incredibly complex task requiring full LCAs, analysis of how a vehicle is used and maintained, its lifespan, and how it is disposed of, to name but a few. This is before we consider whether it crashed into something or took up too much road space.

Instead, we are proposing using mass as a *single, easy metric* to approximate those environmental effects. We should not let the perfect be the enemy of the good, especially when we are facing urgent environmental challenges. By its very nature, the system is not perfectly accurate—as we are prioritizing usability of perfection—so it is important to understand its limitations, and whether it creates any noteworthy secondary effects.

Does This Not Just Incentivize ICEVs?

The first issue, and perhaps most obvious, is that on average fossil-fueled ICEVs are *lighter* than BEVs but have *greater* tailpipe and often life cycle carbon dioxide emissions. Have we not just created a new flawed system that acts in favor of polluting combustion engine vehicles? This would be true if we abolished taxation on fossil fuels (gasoline and diesel primarily). We have talked about replacing the falling revenues associated with it because as more people buy BEVs the number of people buying gasoline and diesel will decrease, but we *do not* propose abolishing it immediately, or indeed at all.

One of the main aims of our scheme is properly to incentivize real reductions in greenhouse gas emissions, mainly carbon dioxide. Our scheme does that, with so many parts of the emissions impact of vehicles being linked to their weight *within* ICEVs and *within* BEVs, but not *between* these two groups. Therefore, liquid fuel taxation still has an important role in making the weight dependency work *between* the groups.

Let us stick to our UK example to apply this. Chapter 12 saw us calculate the tax rates under our proposed weight-based system, assuming all vehicle taxes were replaced and the level of total tax revenues maintained constant. If we now retain the existing fuel duty as a fossil tax, that would lead to additional revenues and so breach the revenue-neutrality goal, until purchases of fossil fuels have reduced to immaterial levels. But there is an additional point. Although many people and politicians see fuel duty as being at a high level today, that level is not high enough to price the carbon emissions completely. Therefore, to achieve a new universal tax system, which is technology neutral between ICEVs, BEVs and other technologies, the rate of fuel duty in fact has to increase. For clarity, non-fossil liquid fuels including e-fuels would not be subject to the fuel duty. If constructed in the following way, we believe it is possible to achieve a tax system that is both technology- and revenue-neutral during the energy transition.

Let us consider an example D-segment vehicle (a fairly large family car) to illustrate: the Volkswagen Passat is a popular choice in the UK and weighs around 1600 kg (3500 lb), depending on optional extras [13.1], and costs between £22,240 to £42,590 ($28,000-$54,000) [13.2]. It is no longer manufactured, as it was phased out in 2023 having been available since 1973 [13.3]. Its rough replacement is the Volkswagen ID.7, a BEV. This weighs around 2250 kg (5000 lb) and costs start at £55,570 ($70,000) [13.4, 13.5]. They are, broadly, equivalent cars (see Figures 13.1 and 13.2 to judge for yourself) and the Passat is in line with the average weight of the cars in the UK today, therefore these suit our comparison well.

What should the optimal level of fuel duty be? As an average car, the 1600 kg Passat is taxed at 7.7 pence per mile under our proposed weight-based system, whereas, weighing 2250 kg, the ID.7 will face 10.6 pence per mile. As it is the case that over a vehicle lifetime an average BEV has approximately half the carbon dioxide emissions of an average ICEV, this would imply that the ICEV should attract tax of 13.5 pence per mile (10.6 divided by 50% minus 7.7) [13.6, 13.7]. This is 2.5 times the 5.5 pence per mile level for the Passat under the fuel duty regime today. As fuel duty becomes a fossil tax, the primary relevant pollutant is carbon dioxide, so we can confine our calculation to this. For context, this would make fuel duty a hefty £1.55 per liter ($7.44 per US gallon), which would only be payable on fossil fuels not on biofuels or e-fuels.

If we moved to this extremely high level of fuel duty, not only would it have significant regressive effects on taxpayers, but also it would raise significant amounts of additional tax, which would violate the revenue-neutrality objective. Therefore, the solution is to have a phased transition from the current system to one which is almost entirely weight-based, except for a small volume of legacy fossil fuel sales. Revenue neutrality is ensured during this transition if the weighting between the old and new systems reflects the proportion of non-fossil-fuel vehicles on the road. Today, this is 3.5% [13.8]. In year one of the transition, we calculate that the

FIGURE 13.1   VW Passat.

Mateusz Rostek/Shutterstock.com.

FIGURE 13.2   VW ID.7.

© Alexander-93/https://en.wikipedia.org/wiki/Volkswagen_ID.7#/media/File:Volkswagen_ID.7_1X7A1779.jpg.

weight-based formula would be weighted at 23%, reflecting a combination of this 3.5% plus the revenue hole left by the abolition of VED. Fuel duty would also be increased by the 3.5%. As a result, therefore, the tax on the Passat in the first year would be 9.5 pence per mile, which would be a combination of the mass-based tax and fuel duty, compared to 5.2 pence per mile for the ID.7 solely from the mass tax.

As a reference, currently fuel duty works out as a tax of approximately 5.5 pence per mile driven for an average car, on top of which there is £203 VED for an average ICEV and, starting from 2025, VED will apply to BEVs too. Therefore, the first year of the transition system would see an increase in tax paid for the ICEV and a reduction for the BEV. This is a justified, technology-neutral, revenue-neutral, incentive to move to a lower carbon vehicle.

You can clearly see that such a system would incentivize a switch away from fossil fuels toward more sustainable options. We should note, also, that this is perhaps a little harsh on the fossil-fueled ICEV, in that we are only considering the carbon dioxide emissions here, but not the fact that it will tear up the road less or be safer for someone it hits, among other advantages that you will be familiar with from previous chapters. Perhaps we should revise our number down? But, if so, by how much? That is ultimately a choice for politicians, based on the perceived significance of the different pollutants.

You will note that so far we have been careful in this chapter to talk about fossil-fueled ICEVs rather than ICEVs in general. Importantly, it is the fossil nature of the fuel that is the major problem with these vehicles more so than the ICE technology itself. There are alternatives to fossil fuels for these vehicles, particularly biofuels and e-fuels, as leading examples of sustainable fuels. As a brief aside, biofuels are fuels that are made from recent biomass (crops or waste cooking oil as examples) and can have substantially lower or even near-zero carbon dioxide emissions as the emissions produced when burning the fuel are absorbed by the plant that grew to produce it in the first place. E-fuels are similar except that the carbon dioxide capture is not done by a plant, but artificially either by capturing the emissions at a source like a brewery or power plant, or directly from the air. In a brewery, the fizz you get in beer (or champagne—depending on your tastes) comes directly from ethanol fermentation; as such, there is a relatively pure carbon dioxide stream to capture at a brewery, so it is, in relative terms, quite easy to do [13.9]. Similarly, in a fossil-fueled power station, the carbon dioxide stream from the combustion process is mixed with water vapor and so again is relatively easy to capture [13.10]. Direct air capture, given the low concentrations of carbon dioxide in the air, is much more difficult—and expensive [13.11]. A much more comprehensive explanation of both biofuels and e-fuels is given in Felix's previous book with Kelly Senecal, *Racing Toward Zero* [13.6]. These fuels, like any, are not perfect, but they can, and do, substantially reduce carbon dioxide emissions from ICEVs and can do so with the existing fleet of vehicles rather than relying on, and waiting for, people to buy new vehicles. As an example, doubling the content of biofuels in gasoline in the UK in September 2021—moving from up to 5% ethanol to up to 10% ethanol—is estimated to have saved 750,000 tonnes of carbon dioxide emissions per year [13.12]. It makes no sense to disincentivize such large potential savings.

Therefore, fuel taxation should *only* be on fossil fuels and *not* on sustainable fuels. There could easily be a sliding scale for blends of fossil and sustainable fuels. This has an additional benefit, incentivizing the development of sustainable fuels. These sustainable fuels are not, at least not yet, cost-competitive with fossil fuels at scale, but if there were a £1.55 per liter ($7.44 per US gallon—yes, those howls of protest again) tax on fossil fuels, sustainable fuels would be looking highly price-competitive, quickly. It would not be practical politically or commercially to jack up fossil fuel taxes all the way immediately. Nevertheless, these levels are both a useful benchmark of environmental effects—in relative terms to our tax take today, and a path that could be plotted. In the future, once we have sustainable fuels and BEVs adopted in the mass market, if someone really wants to use fossil fuel, they should pay the full economic price, including all environmental externalities.

Significantly increasing the use of sustainable fuels in ICEs has many benefits for rapid decarbonization, and our system incentivizes this strongly. It is important to remember that we are not here to judge technologies or have fights about ICEVs vs BEVs—as we have explained, we are proposing a system for labeling and taxation in a rapidly electrifying world where we can effectively and easily account for the environmental impact of vehicles.

What about Hydrogen?

So far, we have been almost entirely comparing BEVs and ICEVs. These are not the only options, though. Hydrogen is another possible alternative, although how much market share such technologies achieve in the light-duty vehicle market, we think, is questionable. Whether or not hydrogen ever makes it into the mass automotive market, our proposal would hold for these vehicles relatively well. While hydrogen is very light—it has a high gravimetric energy density (energy per unit mass) of just under triple that of gasoline—the equipment required to handle and store it safely is heavy (hydrogen, being the smallest molecule, is prone to leaking and is also highly flammable). In addition to burning it in an ICE, you can also use it in a fuel cell, as has been done in hydrogen cars so far. You will need a relatively heavy fuel-cell stack, a set of radiators to dissipate the heat, and a set of batteries (smaller than in a BEV) to buffer the fuel-cell output.

By far the most popular hydrogen car in the world is the Toyota Mirai. It should be noted that in relative terms, however, this is not at all a popular car—22,000 have been sold in total since 2014 [13.13]. In contrast, the Toyota Corolla sold 1.12 million units in 2022 alone [13.14]. Conveniently, the Mirai sits in the same market segment as the Passat and ID.7 that we looked at previously (judge for yourself in Figure 13.3). The Toyota Mirai weighs 1900 kg (4200 lb)—about halfway between the Passat and ID.7. The environmental impact of FCEVs can vary wildly. Of course, they do not have a tailpipe that emits anything other than water, but their life cycle emissions depend heavily on the carbon intensity of the hydrogen production. If it is produced from coal or natural gas—as 98% of hydrogen is worldwide today (so-called brown or gray hydrogen)— then the footprint can be high [13.6]. If it is produced with carbon capture, utilization, and storage (CCUS) or electrolysis of water, then the footprint can be very low. Therefore, if hydrogen were subject to a pollution tax depending on its provenance in the same way that we propose for fossil gasoline and diesel, then our system would continue to work well.

FIGURE 13.3   Toyota Mirai.

Foto by M/Shutterstock.com.

What about Very Lightweight Materials?

We saw in Chapter 8, when discussing LCA, that lighter-weight materials often have higher carbon footprints. In particular, we noted that the carbon dioxide (equivalent) emitted per tonne of aluminum manufactured is around six times higher than that of steel, but replacement of key steel components with aluminum leads to a weight saving of 20–30%. The carbon dioxide (equivalent) emitted per tonne of carbon fiber manufactured is around 11 times higher than that of steel, and carbon fiber replacement of steel parts can commonly lead to weight savings of approximately 50% [13.15].

However, there are some quite complex interactions here. By reducing the mass of a vehicle, you are reducing the energy that is required to drag it around, and so the in-use phase emissions from these lighter vehicles are lower. How much lower depends on how far they are driven. In addition, there are practical barriers, such as cost, which means that at the levels of taxation we are proposing, it is not going to be beneficial to stop making cars out of steel and start making them out of carbon fiber. Steel costs £600 to £1200 ($750 to $1500) per tonne, aluminum £1600 to £2800 ($2100 to $3500) per tonne, and carbon fiber £24,000 to £67,000 ($30,000 to $85,000) per tonne. Even acknowledging that you will need less mass of them for your lightweight vehicle, this is still going to be a substantial cost increase.

With these numbers in mind, we do not think our system—in practice—will incentivize the shift to more environmentally damaging, but lightweight materials in a significant way. Of course, if these prices reduce substantially in the future, then some modification would be needed to our proposal to account for this.

What about Sports Cars?

Speaking of carbon fiber, it is worth thinking about sports cars. These will be clearly taxed less heavily under our system as they are relatively lightweight and tend to be driven fewer miles. So, yes, they will effectively be incentivized. In many ways, this is no bad thing, especially with continued fuel duty on liquid fossil fuels. They are in general lower emitting and have a low impact on other areas of the environment *because they are light in weight*. We acknowledge that incentivizing these sports cars may cause some political issue as they are recreational rather than utilitarian vehicles causing environmental damage. However, the aim here is not to be morally censorious, or opine on subjective questions—we will leave that to politicians. We put forward our system as an ethically neutral one.

Jevons' Paradox

As vehicles become lighter, and hence more energy- or fuel-efficient, there is a risk that they will be driven more. This is an example of the Jevons' paradox [13.16], although real-world examples of this paradox are less common than its popularity might suggest [13.17]. Throughout history, we have seen that an increase in fuel efficiency paradoxically tends to reduce fuel use by *less* than the increase in the efficiency would suggest. This occurs because the cost decreases so people can afford to use their car, or steam engine, more often. In economic terms, there are both income and substitution effects: the increased efficiency makes you richer in terms of car miles, so you consume more, and it makes car miles cheaper relative to other forms of transport, so you switch from the bus to the car. This is why distance remains a key part of our calculation—use your vehicle more and you will pay more.

   Nevertheless, if you buy a lighter vehicle your tax will go down for the same distance driven, which is the whole idea of the system, but it may make you feel that you can now drive further. To illustrate this, let us assume you are an average Californian car owner, with a 4000 lb (1800 kg) car that you drive 12,524 miles/year (the current averages). Under our system, you would pay a tax of $1465 per year. Suppose you swap to a vehicle 10% lighter. This would reduce the car mass to 3600 lb (1633 kg), which it is worth recalling, as we saw in Chapter 11, would be roughly equivalent to sticking with the same model, but from ten years ago. Your tax, if you drove the same distance, would now be $1318. However, you now feel like you can afford to drive further, so you take the longer, scenic, coastal road, or you drive to visit your aunt a couple more times, or whatever it is because you really can afford $1465 per year or more. Now you drive 13,925 miles/year, because that would bring your tax—and probably, roughly, your fuel (or electricity) costs—back to that same level of $1465 per year.

Clearly, if this happens, this would to some extent reduce the efficacy of the *Molden-Leach Conjecture* and our proposal, although our system has some ways that can prevent this, at least to some extent. A case in point is that our system relies on a target or average for both mass and distance. If people start driving more, policy makers could simply reduce that target mileage number, and the tax would go back up again without removing the weight disincentive. In our Californian example, if you adjust the target mileage for California from 12,524 down to 11,270, then that 10% lighter vehicle would still pay a tax of $1465 per year. In effect, you will have brought down the average mass of your vehicle fleet *and* the average mileage and still get the same tax revenue. This achieves behavior change. In addition, of course, if everyone did this and drove more, the traffic would be much worse and so people may then not drive as much because the traffic is so awful, provided the solution was not to build more roads to alleviate this problem, thereby inducing more demand.

What about Vehicles That Are Not Cars?

This book focuses on cars—the so-called light-duty automotive sector. We have not discussed trucks, vans, and off-road equipment. In part this is because the current systems for taxing them are very different, vary more from country to country, and are more based on mass already. It is also hard to come up with a distance-based tax for off-road machinery, such as a digger or crane, for the obvious reason that plenty of these bits of equipment do not move very far if at all, yet they are performing a lot of work—their energy or fuel is not used to create motion, but to dig holes or lift things.

In these sectors it is a much more open question of what the powertrain mix will be in the future. Will battery electric powertrains dominate in these sectors? This is not clear yet, and this is an area where we may see more powertrain diversity than we expect for light-duty vehicles.

Nevertheless, there is no reason why a mass-based system would not work well for trucks and vans. The underlying theory that we developed in this book—based essentially on a notion of work done—holds just as well for them as it does for cars, even if the use cases and sizes are different. For off-road machinery, we could potentially use the power of the machine multiplied by the hours it is run per year, for example. Alternatively, mass could still be in there as a scaling factor. In short, the key elements of the theory can remain the same, and maybe we will elaborate further in a future book.

Summary

In this chapter, then, we have seen that while there are some limitations to the *Molden-Leach Conjecture*, we think that they are either not material or can be mitigated. Certainly, within vehicle classes, the *Molden-Leach Conjecture* holds, but to differentiate between vehicle classes, fossil fuel taxation will need to be retained in order to balance electric vehicles, ICEV and hydrogen vehicles. There are some other edge effects that might lead to unintended consequences, although we consider them to be sufficiently unlikely as not to undermine the *Molden-Leach Conjecture*. Our proposal is not perfect, but its simplicity is a substantial improvement on the status quo and it is easily good enough.

14

Efficiency and Misconceptions

IN THIS CHAPTER

- Efficiency is a common metric used to assess vehicles, but there are many divergent concepts of efficiency.

- Efficiency does not generally correlate well with environmental impact.

- We consider how an optimal balance between car usage and other modes could be assessed.

One of the great claims in favor of BEVs is how efficient they are. The level of efficiency is often claimed to be in excess of 90%, which relies on the concept of how much of the energy stored in the battery is turned to motion of the vehicle. But this is a very narrow and unrepresentative concept of efficiency. It is not even a direct measure of environmental impact, but relies upon the indirect link to how carbon and other pollutants are released as the energy is consumed or fuel burnt. In this chapter, we want to examine different concepts of efficiency and what they can tell us about the environmental credentials of vehicles.

Efficiency is a word that we all use, and like to use, a lot; we all have things we want to do more efficiently. Similarly, when it comes to our cars, for more than 50 years we have been trying to make them more efficient. However, what does efficiency actually mean? Why are we interested in it? *The Oxford English Dictionary* defines efficiency as both "Fitness or power to accomplish, or success in accomplishing, the purpose intended" and "The ratio of useful

work performed to the total energy expended or heat taken in."
[14.1]. It is perhaps this dual definition, and therefore the mixture
of these two quite contrasting things, that leads to our perceptions
that we have of efficiency today.

Generically, "efficiency" is about how much output you can get
for a given input. The more you get, the more efficient the process
is. The concept of efficiency described above is energy efficiency;
the equivalent for ICEVs is fuel efficiency—miles per gallon in the
UK and the US—which is how many miles (the output) can
be achieved for a unit of fuel (the input). In most other countries,
fuel efficiency is not used in common parlance, but rather they use
fuel consumption, which is the inverse: how much fuel (the input)
is required to go a given distance (the output), often using the liters
per 100 km unit. Fuel consumption is not strictly a measure of
efficiency, but it is perfectly inversely correlated to fuel efficiency.

When considering efficiency, there are two areas to consider.
First, what are the relevant input and output variables? Second,
what are the "system boundaries."

The inputs and outputs can be chosen in many combinations.
Energy or fuel efficiency is not the only choice when trying to
characterize the environmental aspect of vehicles. Rather than
relying upon the indirect nature of energy efficiency, we can make
the efficiency directly linked to, for example, carbon dioxide. The
efficiency would then be "carbon efficiency," that is, how far a car
could travel for a unit of carbon dioxide released. Then there is
"economic efficiency," that is, how far a car can go for one pound
or dollar of fuel—although we must be careful not to confuse this
with completely separate notions of economic efficiency.

Following the chain of thought of the *Molden-Leach Conjecture*,
we will coin a new concept of efficiency: "mass efficiency." This is
how much transportation utility you can get for a certain mass of
vehicle. This is a proxy for carbon or environmental efficiency, but
much easier to estimate and integrate into your thinking.

When considering any of these, it is also important to set our
"system boundary." Where do we consider the inputs and outputs

for our efficiency calculation? Is it just the car? The whole system it ever directly or indirectly interacts with? The planet? Just our use of the car? Do we go back to the plug or gas (petrol) station? What about back to the oil well or wind turbine? Any of these choices could drastically affect the calculated result that gets quoted as "efficiency." We would also note that, as a result, efficiency statistics can be widely misused. It is tempting just to think about the vehicle, which is the object that we see and interact with, but it is also the part we pay for directly. But, in doing so, we can miss the bigger picture, for example the upstream emissions, which, as we have seen in this book, are important. By putting too close a focus on vehicle-level efficiency, we can encourage suboptimal behavior outside of that system—and even end up with incentives that are harmful.

Energy Efficiency

Let us start by considering energy or fuel efficiency. Noting the points about where to draw our system boundary, on one level, energy efficiency is relatively simple when thinking about cars. When teaching undergraduate courses, when describing efficiency, the phrase often used is "what you get divided by what you pay for," or the output divided by the input, and that can quite easily be applied here. It can be summarized as "What proportion of the energy you input gets converted into the motive force that moves the vehicle forward?" However, even if we just consider vehicle-level energy efficiency, that too will vary significantly—for starters, if you are stationary with the engine running, by one definition it will be zero (you are not going anywhere—therefore, no motive force moving the vehicle, or no miles whatever the number of gallons).

We know that both ICEs and electric motors have performance and hence energy efficiencies that vary with how fast they are going and the load (force demand) that is required of them. Hence, as a vehicle drives around at different speeds and with different load demands—how hard you accelerate, the number of people on board, the amount of stuff in the boot (trunk), whether you are going uphill or downhill, and so on—the performance of the engine or motor will vary substantially and so the overall energy efficiency of the vehicle will vary. What do we do? Take an average? That, often, is the only way out of this quandary, but at the very least we should acknowledge the variability possible around such an average even for the same vehicle, going the same route, with the same amount of stuff on board but being driven by a different person. Alternatively, we take the person out of it and have all cars be driven by robots (autonomously, in the jargon), but that is a topic for another book. A more useful option, perhaps, is to express efficiencies within ranges. These ranges need not necessarily be the absolute maximum and minimum possible (as we have seen above, the minimum efficiency will always be zero) but plausible representative ranges for common operation.

When comparing the efficiency of BEVs and ICEVs, BEVs stand out as significantly more efficient at a vehicle level. They convert 70–85% of electrical energy from the grid into energy delivered at the wheels, even when you take into account charging losses [14.2]. In contrast, a conventional ICEV only transforms approximately 12–35% of the energy stored in gasoline into wheel power, which increases to 21–40% for a full hybrid [14.3]. Both BEVs and hybrids also benefit from regenerative braking, recovering energy that would otherwise be lost to heat through friction as the car brakes— by using the internal friction of the electric motors to help slow the car down, some electricity is captured, which is then fed back to charge the battery [14.4].

So far, we have been talking about a system boundary around the vehicle only. We know that this is not enough. To offer a comprehensive perspective, we need to go back to the energy or fuel source, recognizing that batteries and gasoline tanks merely store energy and do not generate it. Gasoline and diesel need to be refined and transported to the pump; this has an efficiency of 85–95% [14.5].

For hydrogen, assuming that you store it at 700 bar on a vehicle (the most common way), the compression of hydrogen to this pressure is approximately 85% efficient [14.6]. Beyond that, the efficiency depends on where you get the hydrogen from—if it is from coal (approximately 27% of global production), oil (approximately 22%), or natural gas (approximately 47%), then its production is 65–75% efficient [14.7]. If, as most people desire, because it is the route to near-zero-carbon fuel, the hydrogen is made by electrolysis of water (approximately 4% of global production), then it is 70–80% efficient, although this does not include the efficiency of the electricity generation, which needs to be considered next [14.8].

Electricity generation has a wide array of different efficiencies depending on where you get it from. Combined cycle gas turbines (CCGTs) are the most efficient at 50–64%, wind is 45–48% efficient, coal/oil/gas power plants (not combined cycle) are 37–46% efficient, and solar is 15–27% efficient [14.2, 14.9–14.14]. Additionally, the transmission and distribution of electricity result in an additional 5% loss in the US, and this figure is reasonably representative of many countries [14.15].

Biofuels have poor efficiencies, primarily because photosynthesis in plants is very inefficient at converting incident solar energy into useful energy for the plant (0.1–10% depending on the plant) [14.16]. Once you add in production of the biofuel (for example, ethanol) from the plant stock, overall efficiencies are in the ballpark of 0.1% [14.17]. For e-fuels, the efficiency is also low because there are a lot of steps (carbon dioxide capture, hydrogen generation, and then fuel synthesis)—although nowhere near as low as photosynthesis. Power-to-fuel efficiency (how much fuel you get for unit input energy) is of the order of 30–60% [14.18, 14.19]. On top of that, you would still have the electrical generation efficiency to factor in—as before.

You can see that we now have ranges for both the efficiencies of the vehicle and their energy sources. Nevertheless, we are still only considering the use-phase efficiency—we have not touched on the life cycle aspects that might affect efficiency. When we combine these vehicle and energy source efficiencies, we get the results shown in Table 14.1.

TABLE 14.1 Summary of vehicle and system efficiencies for light-duty propulsion technologies [14.2, 14.9–14.19].

	Vehicle efficiency (%)	Vehicle + Energy source efficiency (%)
BEV (CCGT)	70–85	33–52
BEV (solar)	70–85	10–23
BEV (steam cycle (coal/oil/gas))	70–85	25–37
BEV (wind)	70–85	32–38
FCEV (SMR hydrogen)	40–60	22–38
FCEV (electrolytic hydrogen)	40–60	4–24
Full hybrid (biogasoline)	21–40	0.02
Full hybrid (e-gasoline–solar)	21–40	0.95–6.5
Full hybrid (e-gasoline–wind)	21–40	2.8–11.5
Full hybrid (gasoline)	21–40	18–38
Hydrogen ICEV (SMR hydrogen)	14–43	8–28
Hydrogen ICEV (electrolytic hydrogen)	14–43	1–20
ICEV (biodiesel)	12–45	0.01
ICEV (biogasoline)	12–35	0.01
ICEV (e-diesel–solar)	12–45	0.5–7.3
ICEV (e-gasoline–solar)	12–35	0.5–5.7
ICEV (e-diesel–wind)	12–45	1.6–13
ICEV (e-gasoline–wind)	12–35	1.6–10
ICEV (fossil diesel)	12–45	10–43
ICEV (fossil gasoline)	12–35	10–33
Plug-in hybrid (biogasoline)	21–85	0.02–52
Plug-in hybrid (e-gasoline–solar)	21–85	0.95–52
Plug-in hybrid (e-gasoline–wind)	21–85	2.8–52
Plug-in hybrid (fossil gasoline)	21–85	18–52

© SAE International.

This is an important table, as it shows no one option stands out, although the efficiencies of hydrogen powertrains are clearly in a lower bracket, and e-fuels, and particularly biofuels, have very low energy efficiencies. This clearly shows that claiming that BEVs are up to 90% efficient in turning energy stored in the battery to motion is not incorrect, but it is largely irrelevant. When you take into account the inefficiencies within wider system boundaries, you see that they have almost no energy efficiency advantage. But further, the similarity of these results for vehicles that we know have different environmental effects shows that energy efficiency is a poor proxy for pollution. Of course, energy efficiency is related to these things, but the link is far from direct. Do we care that solar panels are so inefficient in capturing the abundant free energy from the sun that hits the earth? Do we care that plants are so inefficient (and make solar panels look brilliant at converting the energy from the sun)? They are environmentally beneficial even though they are inefficient, as defined. Therefore, we do not think that energy efficiency is a particularly helpful metric for comparing technologies or making effective decisions in this area.

Carbon Efficiency

Carbon efficiency has the potential to be much more relevant to and aligned with our aims. Put simply, it measures how much carbon dioxide is emitted, on a life cycle basis, per unit of useful motive power, with distance as a useful proxy for this. Here, then, the carbon input, rather than any conversion, is the determining feature in the efficiency. After all, if your input is zero carbon, no matter what you do with it or how you convert it, you will still have a zero-carbon output.

In reality, there is no such thing as a zero-carbon energy input yet. All current energy sources, even renewables, have some carbon associated with them, so this brings us back to our LCA, discussed previously in Chapter 8. What we need to do is to account for the total emissions of carbon dioxide associated with the vehicle and spread them over the expected usage of the vehicle. As we noted back in that chapter, it is easier said than done. Nevertheless, we consider this to be a much more appropriate way to think about "efficiency" in vehicles. It is also important to remember that carbon dioxide is not the only greenhouse gas. Others such as methane and nitrous oxide are also extremely potent. Today, they are not emitted in substantial enough quantities from cars to warrant much attention, but we must make sure that in a focus on "carbon efficiency" we do not inadvertently raise emissions of these other greenhouse gases. Perhaps this measure, in time, will become "greenhouse effect efficiency."

Economic Efficiency

Economic efficiency, in strict economic terms, comes in many flavors including allocative and Pareto efficiency [14.20, 14.21]. They express a notion of optimality as to the distribution and application of scarce resource. We take the same idea and stretch it to mean the amount of transportation service you can get per unit of money spent. It is really a notion of cost efficiency of motoring. For example, a fuel may burn energy-inefficiently in an engine, but if it is cheap to buy then it may be economically efficient. We need to be careful in making such assessments, however, as there is much more to the consumer's willingness to pay than the functional utility delivered. Beyond transporting you, branding and design are valued highly in cars, and they form a status symbol. Therefore, the "transportation service" is not just the miles driven but the "quality" of that experience as well.

On the cost side, there is much more to it than the price of the fuel. To understand the total costs (the inputs), we need to include the upfront costs of a vehicle purchase as well as the operating costs—sometimes referred to as the total cost of ownership (TCO). Both factor into economic efficiency. Consumers need to be able to afford to buy the vehicle in the first place (whether new or secondhand) even if the operating costs would end up working in their favor in the long term. People are often rather bad at accounting for these future costs in their decision-making, which is known as applying too high a discount rate [14.22]. As vehicle purchases move increasingly to a financed or leasing model, this problem diminishes as the costs are spread broadly in line with the benefits.

Mass Efficiency

Mass efficiency, finally, is what we are really driving at in this book. Effectively, it is how heavy a car is relative to the utility of that vehicle. The mass part of that equation is easy; the utility part is a little more subjective. Utility needs to balance practical consider-ations (such as number of seats and baggage capacity), range or charging time constraints, safety, and other factors such as off-road and towing capability. In reality, however, we know that most cars meet most utility requirements—even very small cars such as Japanese kei cars can transport two people comfortably, and the average vehicle occupancy in both Europe and the US is 1.5–1.7 [14.23, 14.24]. Yet, people often do not buy these vehicles. Part of the reason can be put down to branding, and part to the "what if" contingency—I do not need these features now, but I might in the future.

Clearly, from the discussions in this chapter we can see that these efficiencies do not always align. Energy efficiency has such a wide range of outcomes as to be unhelpful; carbon efficiency will prioritize anything with a zero-carbon input (e.g., hydrogen, e-fuel, or battery electric energy from solar or wind), but at the moment it is very difficult to have certainty about maintaining that zero-carbon input, as, for example, you cannot only charge a BEV with zero-carbon electricity or use hydrogen exclusively made by electrolysis. Economic efficiency is likely to favor older technology, where development costs and fuel costs have been amortized and are long in the past—such as fossil fuel-powered ICEs. Mass efficiency favors the smallest cars—perhaps even slightly impractical sports cars.

What Would a New Equilibrium of Car Usage and Welfare Look Like?

Based on the underlying wish from this book to tackle the environmental pollution from vehicles in an optimal way, we have proposed our new labeling and taxation mechanism. To end this chapter, it is worth briefly considering how we weigh up car usage and human welfare. Getting out our *Oxford English Dictionary* again, welfare is: "The state or condition of doing or being well; well-being, prosperity, success; the health, happiness, and fortunes of a person or group." Cars bring immense benefits to people, advance human mobility (we like to travel), and offer both perceived and real independence. This is not without its costs, environmental and otherwise, as we have seen. How can we think about the balance between these benefits and costs?

Achieving a new equilibrium of car ownership, usage, and welfare involves balancing the benefits of transportation with considerations for environmental sustainability, public health, and social well-being. Our proposal points the way. There are two generic options: drive a lighter car and/or drive fewer miles. The former is a purchase decision, and—in a sense—not buying a car at all is buying an infinitely light car. The latter option of driving fewer miles manifests itself in many ways today.

Sustainable Transportation

Promote and invest in attractive public transportation options. This would include buses, trains, (including high-speed rail between cities) and underground metro systems (subways). Alongside this, promote and invest in active travel options, including cycling and walking, which are true zero-carbon personal mobility.

Urban Planning

Cities and towns could be designed with mixed-use developments, which reduces the need for long commutes and promotes walking. Urban planning could also consider connectivity of new developments—not leaving them reliant on cars for mobility. Green spaces in developments could be essential to tempt people to spend leisure time there rather than traveling elsewhere.

Incentives for Carpooling and Ride-Sharing

Incentives could be offered for carpooling and ride-sharing services to reduce the number of single-occupancy vehicles on the road. Such incentives may be traffic based, for example, high-occupancy vehicle (HOV) lanes, financial, or technological, such as Uber or Lyft for carpooling.

Flexible Work and School Arrangements

The problem with any infrastructure system is that it can be hard to handle fluctuations in demand, as that infrastructure tends to have a fixed or near-fixed capacity. With our levels of population and economic activity today, we often reach that capacity in transportation, especially at "rush hour," a peak in the morning and evening that often lasts more than an hour. If we increase flexible work arrangements, such as remote work or flexible hours, this not only reduces the peak use of infrastructure but may also reduce the need for daily commuting overall. Many have tried this as a result of the COVID-19 pandemic, with varying success, but businesses now have a much clearer idea of which telecommuting options and virtual collaboration tools work well for them. Not all interventions are good, however, as shown by the abolition of the universal weekend in early Soviet Russia, which was a complete failure and reversed. Some variability is clearly advantageous in spreading the load on our infrastructure to the benefit of everyone [14.25]. Cities see this during school holidays when traffic is lower, and prices of holidays are higher.

Education and Awareness

Societal change often comes with education. We can raise awareness about the environmental and health impacts of excessive car usage—particularly excessive, large car usage. By understanding the environmental effects, there might be a move to "right size" both our cars and our usage of them. Perhaps as part of this education and awareness campaign, you could recommend a friend buy this book.

Summary

While all of these sound meritorious, to assess them properly you need to think it not just in terms of our mass × distance concept, because that says nothing about the utility side of the equation. Pollution could be reduced by avoiding car driving and switching to active transport or not traveling at all. But what is the utility change, whether positive or negative? Ideally, we could travel less and be happier—that is an easy decision. More often, reducing distance reduces utility and we have to compromise on some life-style or work element. It is not the job of this book to determine these complex and subjective utility questions, but it does give you—through the metrics of efficiency—a way to assess if you are traveling in the best way, and, through the *Molden-Leach Conjecture*, what the overall environmental impact of your behavior is.

15

Hidden Dysfunction

IN THIS CHAPTER

- There are three approaches to limiting bad things: ban them, limit them, or tax them.
- Current regulatory systems are flawed and even incentivize larger vehicles.
- Europe has a patchier track record than the US.

This book is not simply putting forward an approach to vehicle labeling to enable good consumer decisions and effective government taxation. It offers, in a world of rapidly electrifying transport, a reset. Vehicle emissions regulation, which started in the 1950s in California, but developed new urgency in the 1970s, has been successful in many ways. The gains have, over time, become smaller and less certain; the existing regulatory regimes may soon have outlasted their usefulness. To understand this, we should look at the troubled recent history of emissions regulation in the US and Europe.

Europe

Not usually of great consumer interest, emissions hit the headlines in September 2015 with the "Dieselgate scandal," when Volkswagen was found to have broken the US certification rules by deploying "defeat devices" that kept emissions artificially low during the official test to achieve a pass and be legal to sell [15.2]. The real significance of this event is not as it seems. It was not that one manufacturer had cheated and suffered large fines, restitution, and criminal convictions for some of its employees. It was more about what it reflected about limit-value systems worldwide.

Certifying vehicles to meet all the necessary limit values is an expensive process. There is the primary cost of engineering the vehicle actually to meet the limits, which typically involves a large development investment that can easily add thousands of dollars to the purchase price of a vehicle. The certification process itself requires conducting tests and submitting extensive documentation to authorities. The testing itself falls into three main areas. First, a vehicle that represents a "type"—or generic group of vehicles of similar mechanical characteristics—must pass the test. Second, to ensure manufacturing is consistent with the performance of the type approved, individual vehicles are randomly selected from the production line and retested. Third, a certain number of aged vehicles must be retested to show they still comply with a factor for potential deterioration [15.9]. Much of the costs of these processes fall on the vehicle manufacturer [15.10]. In addition, the authorities should undertake surveillance to check for the health of the system and pick up any cheating. Any miscreants should suffer enforcement action. All of these come at either a cost to the taxpayer or as fees passed on to the manufacturer. Whoever pays, these are real costs, and the lower the better for consumers and taxpayers. Therefore, a good system is not just one that reduces emissions, but is mindful of the compliance rate (i.e., lack of cheating) and cost of operation.

Ultimately, the real compliance rate is unknown because we do not know what cheating was not caught. What we do know is that the US system has regularly discovered various dubious behavior, including the original "consent" decree settlement with heavy-duty truck manufacturers in 1998 [15.11]. In this case, simple switches were connected to the hood of the vehicle, to see if the hood was up, which was the condition in which certification testing took place. The US EPA conducts more than 100 randomly selected tests on private vehicles each year, which helps it detect infractions, although this is a small number relative to the size of the market and cars on the road. Such testing failed to pick up the cheating by Volkswagen in the run-up to 2015. Instead, it was found by the combination of shrewd investigation and a portion of good luck by a test project commissioned by the International Council for Clean Transportation and conducted by West Virginia University [15.12]. Over a similar timeframe, European authorities discovered nothing. Because they did not look [15.13]. There was no requirement for surveillance until the changes were enacted after Dieselgate.

Although the ground zero of Dieselgate was Volkswagen's violation of the US certification system, the greater story was that every manufacturer was manipulating the certification of their European vehicles. Prior to tightened regulations after Dieselgate, Emissions Analytics had tested 552 European diesel vehicles and found that real-world emissions were on average 4.1 times higher than the emissions limit for nitrogen oxides. Exceedances were seen for all 32 vehicle manufacturers. Some of these manufacturers were selling diesel vehicles into the US market at the same time and achieving the lower emissions required by US regulations. At the time of writing, many legal cases were proceeding through the UK and other European courts, or were being prepared, that will ultimately determine whether these exceedances represented illegal or legal cheating. Whatever the answer, it is worth remembering that, until Volkswagen's violation in the US was uncovered, no authority in Europe was questioning the robustness of their certification system.

It is clear, therefore, that these emissions certification systems are costly and prone to failure, which reflects their complexity. That complexity was arguably needed when tailpipe emissions were the dominant source of environmental contamination from vehicles, and each needed limiting individually for good reason. Now that we have moved to a world of limited tailpipe emissions, what we have shown in the foregoing chapters is that mass is a good proxy for many of them. We can, therefore, set aside the costs and compliance problems of emissions limit values and use a simple alternative variable. The beauty of the vehicle mass proxy is that it is very cheap to observe—a set of vehicle scales is sufficient—and very cheap to verify. No manufacturer would risk falsifying the official kerb weight of their vehicle, as it would be so easy to detect.

An added benefit of regulation and taxation by mass is that it makes it hard for governments to use the regulatory system for directly covert subsidies that may distort the market. The current limit-value-based system in Europe indeed hides several subsidies, even though they are not called by such a term. Most notably, the Euro system has, and this will continue with the introduction of the new Euro 7 rules from 2025, differential limits for gasoline and diesel engines [15.14]. The carbon monoxide limit is higher for gasoline vehicles than it is for diesels, in the same way that the nitrogen oxide limit is higher for diesels compared to gasoline engines. As the nitrogen oxide limit is harder for a diesel engine to meet than the carbon monoxide limit is for a gasoline engine, this confers a net subsidy of the diesel engine. In other words, if the diesel engine had to meet the same limits as the gasoline rival, it would have to deploy more, costly aftertreatment technology. It is worth pausing here, a moment, to consider this. As a breather of air, do you care what emitted the pollutant that you are breathing? Does it matter whether it was a gasoline or diesel vehicle? From a health perspective—the primary driver of these regulations in the first place—it makes absolutely no sense to differentiate by technology type.

A more egregious subsidy for a technology, however, happens through a parallel piece of regulation to the Euro standards—the fleet average carbon dioxide targets [15.15]. Each manufacturer in Europe must, each calendar year, meet an average carbon dioxide emissions value, weighted by sales across its range of vehicles. The carbon dioxide value for each model is based on the official laboratory certification value. Based on the old NEDC test, the headline fleet average target through to 2024 is 95 g/km [15.16]. On the surface, this looks like a reasonably simple and sensible system to chart a route to emissions reduction that gives the industry forward visibility on what they must engineer. However, in practice the system is persistently manipulated by authorities to push a particular technology or advantage a set of interests. This matters because reducing carbon emissions from a vehicle is difficult and expensive. Each one-gram reduction in average emissions requires engineering advances and commercial choices.

The first flaw is caused by the shortcomings of the NEDC test system, as discussed previously, which significantly underestimates, by approximately 30%, the real-world carbon dioxide emissions [15.17]. The WLTP system that eventually replaced the NEDC still sees gaps of around 10% [15.7]. Without a surveillance or correction mechanism to remove systematic gaps, it is in the interests of vehicle manufacturers to manipulate their certification tests as much as possible within the law to get the lowest NEDC carbon dioxide values, even if they are not reflected in real-world driving. While the US has a surveillance system to police these gaps, the EU is only belatedly trying to track this by legislating that vehicle onboard computers will be obliged to relay their fuel consumption—which correlates almost perfectly with carbon dioxide emissions—to a central database [15.18].

A second flaw, deliberately engineered in, is that the 95 g/km target is not a universal one. In fact, each manufacturer gets its own personalized target, based on the nature of their product portfolios [15.16]. In short, if you only sell big cars, that is seen as unfair, so you are given an easier target. There is some logic to this: as bigger vehicles—as we have shown in previous chapters—have inherently higher emissions on average, it would be harder for a manufacturer of these bigger cars to comply with the fleet average target by virtue of the preexisting nature of its business. BMW targets a premium segment with large vehicles and has never offered small, city cars. The flaw with this approach, however, is that it cements in higher emissions from the bigger vehicles. It gives BMW an incentive to reduce emissions *by means other than making vehicles smaller*; otherwise, their target would get harder (that is, they would get a lower fleet average carbon dioxide target). Therefore, in a relatively small but discernible way, this gives an incentive for bigger vehicles as the "cost" to manufacturers of a unit of carbon dioxide from a bigger vehicle is lower than from a smaller car, despite the environmental damage being identical.

By far the most profound flaw in the system is the treatment of BEVs. They count for zero—in a good way. As a pure BEV has no tailpipe, it is designated as "zero emission," which is very useful in dragging down a manufacturer's fleet average toward the target [15.2]. No allowance is given for carbon dioxide emissions shifted up the supply chain to the battery manufacture. And it has been even more extreme than being counted in the average as zero. Between 2020 and 2022, a "supercredits" system was operated [15.19], whereby an BEV initially counted for double: double zero! The level of the credit reduced over time, and was subject to a cap, but nevertheless, it gave a significant incentive toward BEVs while hiding carbon dioxide emissions upstream. In the same vein, between 2025 and 2029 there will be a "zero and low-emission vehicle" (ZLEV) credit system whereby for every percentage point of ZLEV vehicles sold above a 25% target each year, that manufacturer's fleet average target would be increased by one percentage point, up to a maximum of five [15.19]. Together, these constitute a significant incentive to sell BEVs—which are heavier on average than a similar combustion engine vehicle—by a combination of explicit and implicit incentives. The measure of emissions is misleading, and BEVs get a bonus by virtue of their technology rather than their actual emissions.

The net result of all these flaws, especially the second and third ones, is to incentivize selling larger, heavier cars. When we wonder why the average size of vehicles on the road is increasing, it is not a complete answer to say it is shifting consumer preferences. It is shifting preferences, reinforced by embedded incentives in the official European regulatory system.

The objection to this would be to say that BEVs do not have to be that heavy, and if only manufacturers would produce smaller town cars, deploy higher-energy-density solid-state batteries, or lightweight the vehicle body with carbon fiber. While these are all theoretically possible, they are not a practical reality, in part because of the very incentives described above. Smaller town cars are less profitable than high-price SUVs. Solid-state batteries may become commercially viable at mass scale, but that will require significant and continued investment, and currently remain more expensive than standard lithium-ion batteries. In fact, the current trend is toward cheaper, heavier, less energy-dense lithium iron phosphate batteries. Carbon fiber bodies are indeed lighter, but an expensive luxury. Therefore, most manufacturer incentives point toward vehicles of great weight—and this is reflected in the cars on the road today. If we are to reverse this trend, which, as we have shown in the previous chapters, will in most dimensions correlate with lower emissions, the incentives must be shifted. This can be done by a combination of removing the inbuilt indirect incentives toward heavier vehicles and instigating a mass-based labeling and taxation system.

US

Compared to Europe, the US situation shares some of the same downsides, but tends to be manipulated in a less willful way. The emissions certification regime for nitrogen oxides, particulate matter, and other pollutants is in fact structurally more complex than in Europe [15.2]. However, there are three clear advantages. First, the US system is fundamentally technology-neutral—limit values do not differ between gasoline and diesel, for example. Second, the US EPA has a long-established surveillance procedure to try to detect miscreants. Third, instead of fleet-average carbon dioxide targets, the same essential idea is deployed as a CAFE target [15.20]. As fuel economy and carbon dioxide emissions are well correlated for a given fuel, and most cars in the US are gasoline, this makes the system much easier for consumers to understand and engage [15.18]. Consumers do put gasoline into their vehicles, but never measure the carbon dioxide coming out of them. Despite these positives, the US system shares some flaws with Europe, including a "footprint" adjustment in its targets, which advantages bigger-dimension vehicles. This greater size again correlates with greater mass, and this bias is structurally embedded in the regulation. Beyond the CAFE target, the main tool used by some states in the US is a ZEV sales target for each manufacturer each year. For example, California requires 68% of vehicles sold in 2030 to be either pure electric or plug-in hybrids, a milestone on a path to

100% penetration in 2035 [15.21]. While it is effectively creating a bias toward heavier vehicles in practice—although not in principle—it is at least a transparent policy that people can understand and engage with democratically. Unlike Europe, the number of secreted biases in the US is smaller.

In high-level terms, therefore, we can see that emissions regulations have at best had mixed success because of a combination of complexity and deliberate debasement from authorities, governments, and industry, especially in Europe. These two factors also interact with one another, because complexity helps disguise acts of debasement. Like any currency, if you debase it, it ceases to inform actors as to the true situation and, as a result, leads to bad decisions and outcomes. In an effective world, emissions tests should produce results that are a good approximation to what goes on, on average, in the real world. Policies and targets can then be made with reference to the results of those tests—the emissions "currency." These policies and targets are ultimately a political decision—as they both should and can only be—but the emissions currency should be a result of good science, and absolutely not a result of policy.

Summary

Dieselgate was a product of the debasement of the system. The test cycle on which most emissions figures were produced bore little relation to typical driving. The rules then afforded manufacturers the opportunity to manipulate these results—whether this ultimately proves to have been illegal or legal is immaterial. Inevitably, such debasement will destabilize a system to the point of crises, as we saw in September 2015, and the legal cases will rumble through courts for many years to come. Rather than fixing the system, the reaction to Dieselgate was to debase the system even further. Although the inadequate NEDC was replaced by the less-bad WLTP, a much bigger distortion was introduced, in the form of using the system for a purpose it was never designed for. That purpose was to covertly encourage and subsidize BEVs, by giving them a score of zero carbon dioxide emissions. It is, of course, true that these vehicles have no tailpipe and therefore no tailpipe emissions. However, it is illegitimate to use a system designed fundamentally around ICEVs to judge a different technology. It would be as if you regulated the potential environmental fallout from a nuclear plant by using standards for coal power stations. By applying an inappropriate system to regulate BEVs, a wedge is being driven between the reported emissions and reality.

Until the moment the system blows up in a new Dieselgate-type event, many politically attractive statistics for emissions reduction will be created. Due to the complexity of the system, where consumers must in practice trust their authorities and governments, and it is those very bodies which are responsible for the debasement of the system, there will be little awareness of the brewing problem until the day it happens. Remember: Dieselgate came as a bolt out of the blue to everyone except a few experts and researchers, and some politicians and officials who hoped the truth would never surface. There are always plenty of vested interests beyond the authorities to keep the magic going as long as possible, not least industry leaders in electrification who financially benefit from the skewed system.

This current malaise in emissions regulation, and the opportunity and need for simplification afforded by electrification, is exactly the reason we have written this book—to provide a practical alternative that is much easier to run and police, and that will correlate well with the desired political and fiscal outcomes.

16

Where Now?

IN THIS CHAPTER

- Eighty-three percent of unabated pollutants are linked to vehicle mass.

- Taxation, based on the polluter-pays principle, rather than bans, is the better way to move to greener transportation.

- Labeling and taxing cars according to their weight will create the right incentives to correct environmental damage, while also replacing declining vehicle tax revenues.

BEVs are commonly seen as the answer. But to what question? Most people would—in 2024 at least—say carbon dioxide emissions. But as we have seen in this book, such vehicles are far from zero emission. They may be, however, the answer to another, equally important question. The striking property of BEVs is that their emissions correlate extremely well with their mass. So, in a largely battery-powered world, suddenly consumer labeling and taxation become relatively easy. A gift of an easy-to-measure and hard-to-cheat factor that everyone understands! Consumer labeling is revolutionized, in a simple way. Taxation can be based on something that sounds intuitive and fair. Finally, internalizing environmental externalities becomes a practical proposition that would bring about an outcome more economically efficient and socially just.

We still face challenges, however, in the transition because of those pesky ICEVs. But even with these cars, the tailpipe pollutants, except for carbon dioxide, have been strikingly low since around 2018, and carbon dioxide correlates well with vehicle mass. Some of the "clean" combustion vehicles even date back to as far as 2014, although their prevalence in the fleet was patchy. These aside, earlier ICEVs readily spew nitrogen oxides—from diesels—and ultrafine particles—from GDI vehicles—at the tailpipe, with the levels having little to do with the vehicle mass. However, these vehicles are now at least six years old, and many are over ten, so they will progressively disappear from our roads. By the time legislation is passed to reform vehicle taxation and labeling, this disappearance will be even more advanced.

Tax Base

We can, however, embrace that residual complexity to accelerate decarbonization. First, fossil fuel-powered combustion engine vehicles, while heavily taxed, are not taxed enough to account for all the environmental damage that is caused. The contribution to climate change, air pollution, and contamination around drilling and refining sites is great. We should, however, draw a distinction between the "fossil fuel" element more than the "combustion" process. Those fossils lock up carbon sequestered millions of years ago, a process that was necessary for life on Earth to emerge. We cannot risk releasing too much of that. The combustion process is messy, but we have progressively mastered how to clear that up with on-vehicle catalytic converters and filters [16.1]. So, it is the fossil fuel, and its extraction and burning, that is the primary problem, and this must be taxed more [16.2]. By doing this we not only help plug the growing fiscal gap in the switch to electric vehicles, but we also create an incentive to develop non-fossil fuels for combustion engines, such as e-fuels—a market that is struggling to develop as fossil fuel gasoline remains effectively subsidized through its untaxed pollution [16.3].

So, in short: car pollution will increasingly be best approximated by the mass of the vehicle and how far you drive, which simplifies consumer information and presents a fertile new tax base; add to that a proper tax on legacy fossil fuels and you accelerate decarbonization and raise tax revenue in doing so.

In numbers: of the regulated pollutants that have not been effectively abated already, five out of six of them, or 83%, are substantially a function of vehicle mass: in-use carbon dioxide, tire and brake emissions, as well as noise and safety.

And the tax plan: it would start with abolishing existing registration and annual taxes on vehicles, while maintaining or increasing fuel duty, but only on fossil fuels. The new tax would be annual and based on the vehicle weight and the declared miles driven. The level of the tax would start low and progressively increase to replace declining fuel duty on fossil fuels.

Ultimately, placing an average tax of 8 pence (11 US cents) per mile or 5 Euro cents per kilometer for a vehicle of average weight would raise a similar amount in tax to what is currently raised in an average developed country. That tax would then increase by 1% for every percent the vehicle weight is above average, and fall in a similar way for lighter vehicles. Put another way, in the UK, for example, for every 150 kg that can be shaved off the weight of a typical car, it would save the average driver about £100 per year. For 1000 km less distance driven, the saving would also be £100.

Advantages

In one go, you would rebuild the vehicle tax base, simplify it for consumers, and create the best incentives for reducing pollution from vehicles, both in terms of carbon dioxide and environmental contaminants.

Put another way, this seeks to maximize "mass efficiency" rather than energy efficiency. Mass efficiency is getting the maximum amount of motive power, transportation services, or whatever you want to call it, for the minimum mass, because we know mass is the best proxy for environmental effects. Mass efficiency is, therefore, the best proxy for what could loosely be called "environmental efficiency." Environmental efficiency is the amount of environmental benefit that can be achieved for the minimum investment, which could be measured monetarily or via proxies such as mass. And environmental efficiency is the key for humanity to keep enjoying the many benefits of private cars consistent with our need for sustainability and increasingly vexatious environmental constraints. Note also that we do not necessarily maximize energy efficiency. A light, city car powered by wind-derived synthetic gasoline will be very inefficient because of the limitations of both the production and combustion processes, but it will have a low impact on the environment. A big, heavy battery SUV powered by electricity from a natural-gas-fired power station will be relatively energy efficient, but will have a much greater environmental impact.

Our analysis sits within a political context where battery electrification of cars is the consensus, and strongly embedded in legislation, especially in Europe. The story has been sold that battery vehicles solve all problems in one go: the pollution from fossil fuel extraction, refining, and transportation; the carbon dioxide from its combustion; and the air pollutants from the tailpipes of vehicles. Dieselgate gave the perfect moment that coalesced our concerns about climate change, air pollution, and geopolitical dependencies. Between 2016 and 2023, this consensus was dominant, but in 2023 that consensus began to fray as the unintended consequences of battery electrification became more obvious.

Although not the topic of this book, new geopolitical dependencies that sourcing battery materials entail are becoming real, as China starts to ration certain key materials such as refined graphite [16.4]. The environmental and human rights effects of battery material extraction and refining—while sometimes exaggerated by opponents and increasingly mitigated by operators—are real. Then, it becomes clear that the pollutants from the tailpipe of modern vehicles are overall a fraction of those created by non-exhaust pollutants such as tires and brakes.

As a result, we now have an increasing number of policies that have been put in place or are being discussed to mitigate all the ways in which heavy vehicles, and especially BEVs, do not align with the best environmental outcome. The European Commission is running a research program that must deliver by 2025 a potential lifecycle carbon dioxide assessment methodology, with a view potentially to labeling or taxing vehicles according to their manufacturing and other upstream emissions [16.5]. The new Euro 7 vehicle pollution regulation that should come into force from 2025 onward has limit values for the first time on brake and tire emissions [16.6]. There is now an acrimonious wrangle between the European Commission and the German government as to whether low- or zero-carbon synthetic fuels will be allowed to power new combustion engine vehicles after the current legislated ban in 2035—allowing this is a completely logical step if your interests are in carbon reduction rather than blind promotion of battery-powered propulsion. All of these rules will interact in complex and sometimes unpredictable ways, with potentially unintended outcomes. It certainly will be complex for legislators, regulators, and the industry. It will be baffling for consumers.

Our proposal runs contrary, conceptually, to the prevailing political and regulatory orthodoxy in Europe and, to a more limited extent, in the US and other countries. This is sometimes described as a "technology-forcing" approach. Rather than taxing a pollution "bad" and letting people decide how to change their behavior most efficiently, a technology-forcing position is that a problem has been identified and there is a technology that can mitigate it, if only the market can be forced to adopt it. It is either politically or legally wrong to mandate that technology explicitly—perhaps because it will discriminate in favor of some private interest—but it is seen as acceptable to tighten regulations in such a way that manufacturers have no option but to redeploy the technology. Where there are multiple technology options, the legislators are on firmer ground but often, there is a "best available technology" that is known about, and the desire is to force that. A good example of this is the "Euro 5" tailpipe regulators, which tightened the acceptable limit for particle emissions, in a way that forced the adoption of exhaust particle filters on all diesel cars. The limits were sufficiently low that it would be impossible to comply with the law—or, at least, highly risky—without a filter. The result of the regulation was that filters were rolled out quickly and tailpipe particle emissions fell dramatically from diesel cars [16.7]. A success, it would seem.

That same approach is being applied to addressing carbon dioxide emissions. Battery electrification has been identified by legislators—through some combination of research, lobbying, and negotiation—to be the best available technology that should be forced on the car market. The regulators, through carbon dioxide targets, ICEV bans, and ZEV mandates, have created a set of rules that do not specifically mention BEVs *per se*, but these rules are such that they can only be met by BEVs. They cover themselves by pointing to hydrogen fuel-cell vehicles as another option, knowing the technology remains well behind batteries in development such that they are commercially unviable—for now, and perhaps forever.

What is the difference between the case of the particle filters and BEVs? Surely, the filter case shows the power and wisdom of such measures, when environmental and health damage is being caused? The difference is in the complexity. An exhaust filter is a relatively simple, well-understood, proven product, with known production and performance characteristics. Moreover, the deployment of it is relatively straightforward. Manufacturers decide they must install it; they incorporate it in the exhaust line of the next generation of vehicle models and source the necessary units from their supplier of choice. BEVs, in contrast, have incredibly complex, emerging, and often unproven supply chains, under-equipped skills and intellectual property base for research and production, and their real-world production and performance characteristics are still not that well understood. In other words—we are only beginning to understand the downsides, as they are so complex in how they interact with real consumer behavior. The ability to roll them out across the market is also much more complex, in trying to persuade consumers to buy an unfamiliar and often more expensive product. Exhaust filters, in contrast, added a reasonably modest cost to the purchase price, affected the behavior of the vehicle hardly at all, and did not require any consumer behavior change.

Put another way, the benevolent gods sitting in governments and regulators in Europe turn out not to be omnipotent, or indeed omniscient. While they can just about get it right to force a simple device with known properties, they cannot manage it with a product of such complexity as a BEV. And this is completely understandable. No one reasonably could be expected to. It is an alluring policy to choose the technology and mandate it—to pick the winner—because it is simple. But, time and time again, policies collapse when they are complex and collide with the reality of discriminating consumers and real-world behavior. Except for the simplest, most urgent things, they should not, therefore, try.

It is possible, with enough political will and determination, to force any technology through. What bites back—if voters do not—is the real cost. You can force complex technology, but the price of the inefficiency, unknowns, and unintended consequences tends to be very high. What then happens is that governments recoil at the true cost and then ration support for the technology-forcing. The German government did just that in December 2023 when they abruptly withdrew all retail subsidies for BEVs [16.8]. Governments either do not have the depth of pockets or the well of political will to see it through. This is because the original costs were always going to be too high, because the benevolent dictator is fallible and cannot control behavior and allocate resources in an efficient way. We end with half a solution. And half-solutions are often no solution. Often the fallback is setting more and more targets and hoping, via the power of *fait accompli*, that the consumer will see the writing on the wall and adapt their behavior anyway. Rarely does this happen unless consumers see it is in their interest in the first place.

The alternative to this fallacious muddle—which is what we are proposing in this book—is to properly, yet simply and achievably, price the environmental damage. The consumer will then see that price, as a result of the application of the weight/mileage tax, as well as useful accompanying buying information. The consumer can then decide still to buy that car, switch to another one, or not buy anything at all. At the same time, governments will know that their policy is putting that proper price on pollution, so it not only leads to adjusted consumer behavior, but also raises much-needed tax money for transport infrastructure and other policy priorities. Our proposal also has the potential to be a universal one, using mass as a wider proxy for environmental damage. This would help avoid issues highlighted by Lipsey and Lancaster (1956) in their "Theory of Second Best," under which producing the optimal outcome in one area may not be the best overall solution for society if it creates worse distortions elsewhere [16.9].

Summary

Tesla's Cybertruck, perhaps the best emblem for now of the faulty way of thinking our proposal can help solve, is exactly the technology regulators have decided should be forced on us. It is zero emission in law, except it is not in reality. Its construction emissions are very much non-zero, its in-use nonexhaust emissions are material, its impact on infrastructure is predictable, and its safety record will likely prove to be highly asymmetric between driver and pedestrian. Instead, we should move rapidly to a policy not that bans the Cybertruck—this would be an extension of the central dictator fallacy—but to tax it properly in accordance with its environmental impact, which would be on its ample mass.

In this book, therefore, we hope to have argued effectively and shown you, with real data and scientific analysis, that there is a simple, viable way to stop big cars killing us.

References

Chapter 1

1.1. Tetlock, P.E. and Boettger, R., "Accountability Amplifies the Status Quo Effect When Change Creates Victims," *Journal of Behavioral Decision Making* 7, no. 1 (1994): 1-23, doi:https://doi.org/10.1002/bdm.3960070102.

1.2. Ghaffarpasand, O. et al., "Real-World Assessment of Vehicle Air Pollutant Emissions Subset by Vehicle Type, Fuel and EURO Class: New Findings from the Recent UK EDAR Field Campaigns, and Implications for Emissions Restricted Zones," *Science of the Total Environment* 734 (2020): 139416, doi:https://doi.org/10.1016/j.scitotenv.2020.139416.

1.3. Boveroux, F. et al., "Impact of Mileage on Particle Number Emission Factors for EURO5 and EURO6 Diesel Passenger Cars," *Atmospheric Environment* 244 (2021): 117975, doi:https://doi.org/10.1016/j.atmosenv.2020.117975.

1.4. Mock, P., et al., "The WLTP: How a New Test Procedure for Cars Will Affect Fuel Consumption Values in the EU," International Council on Clean Transportation, Vol. 9(3547): 1-20, 2014.

1.5. Valverde, V. et al., "Emission Factors Derived from 13 Euro 6b Light-Duty Vehicles Based on Laboratory and On-Road Measurements," *Atmosphere* 10, no. 5 (2019): 243.

1.6. ICCT, "The Gap between Real-World and Official Values for CO_2 Emissions and Fuel Consumption Grows again Despite New Test Procedure," 2024, accessed February 4, 2024, https://theicct.org/pr-en-gap-between-real-world-and-official-values-for-co2-emissions-and-fuel-consumption-grows-again-despite-new-test-procedure-jan24/.

1.7. Uwe Tietge, S.D., Yang, Z., and Mock, P., "From Laboratory to Road International: A Comparison of Official and Real-World Fuel Consumption and CO_2 Values for Passenger Cars in Europe, the United States, China, and Japan," ICCT, 2017, accessed February 4, 2024, https://theicct.org/publication/from-laboratory-to-road-international-a-comparison-of-official-and-real-world-fuel-consumption-and-co2-values-for-passenger-cars-in-europe-the-united-states-china-and-japan/.

1.8.   Kawamoto, R. et al., "Estimation of CO_2 Emissions of Internal Combustion Engine Vehicle and Battery Electric Vehicle Using LCA," *Sustainability* 11, no. 9 (2019): 2690.

1.9.   Del Pero, F., Delogu, M., and Pierini, M., "Life Cycle Assessment in the Automotive Sector: A Comparative Case Study of Internal Combustion Engine (ICE) and Electric Car," *Procedia Structural Integrity* 12 (2018): 521-537.

1.10.   Randall, A., "The Problem of Market Failure," *Natural Resources Journal* 23, no. 1 (1983): 131-148.

1.11.   Sykes, M. and Axsen, J., "No Free Ride to Zero-Emissions: Simulating a Region's Need to Implement Its Own Zero-Emissions Vehicle (ZEV) Mandate to Achieve 2050 GHG Targets," *Energy Policy* 110 (2017): 447-460.

1.12.   Slowik, P., et al., "Funding the Transition to All Zero-Emission Vehicles," ICCT White Paper, 2019, accessed May 13, 2022, https://theicct.org/wp-content/uploads/2021/06/Funding_transition_ZEV_20191014.pdf.

1.13.   Koetse, M.J. and Hoen, A., "Preferences for Alternative Fuel Vehicles of Company Car Drivers," *Resource and Energy Economics* 37 (2014): 279-301.

1.14.   Liu, Z. et al., "Comparing Total Cost of Ownership of Battery Electric Vehicles and Internal Combustion Engine Vehicles," *Energy Policy* 158 (2021): 112564, doi:https://doi.org/10.1016/j.enpol.2021.112564.

1.15.   Ntombela, M., Musasa, K., and Moloi, K., "A Comprehensive Review for Battery Electric Vehicles (BEV) Drive Circuits Technology, Operations, and Challenges," *World Electric Vehicle Journal* 14, no. 7 (2023): 195.

1.16.   Franco, V. et al., "Real-World Exhaust Emissions from Modern Diesel Cars," ICCT White Paper, October 2014, 1-59, https://theicct.org/sites/default/files/publications/ICCT_PEMS-study_diesel-cars_20141010.pdf.

1.17.   Davison, J. et al., "Gasoline and Diesel Passenger Car Emissions Deterioration Using On-Road Emission Measurements and Measured Mileage," *Atmospheric Environment: X* 14 (2022): 100162.

1.18.   European Environment Agency, "Average Statistics of New Passenger Cars by Member State," accessed February 1, 2024, https://www.eea.europa.eu/data-and-maps/daviz/average-statistics-of-new-passenger-cars.

1.19.   Gasnier, M., "Europe Full Year 2022: Peugeot 208 Ends VW Golf's 14-Year Domination, Market Down –4.1%," 2023, accessed February 15, 2023, https://bestsellingcarsblog.com/2023/02/europe-full-year-2022-peugeot-208-ends-vw-golfs-14-year-domination-market-down-4-1/.

1.20.   Karr, A., "Tesla Model Y Was Europe's Best-Selling New Car Last Year," 2024, accessed January 20, 2024, https://uk.motor1.com/news/705404/tesla-model-y-best-selling-car-europe-2023/.

1.21.   Cars-Data.com, "All Technical Specs in One Car Database," 2024, accessed February 1, 2024, https://www.cars-data.com/en.

1.22.   ultimateSPECS, "Volkswagen Golf 1 1.1 Specs," accessed February 4, 2024, https://www.ultimatespecs.com/car-specs/Volkswagen/2843/Volkswagen-Golf-1-11.html.

1.23. Wallach, O., "The Best-Selling Car in America, Every Year since 1978," 2021, accessed July 16, 2021, https://www.visualcapitalist.com/best-selling-car-in-america-every-year-since-1978/.

1.24. Wayland, M., "A Compact Crossover Is Coming for America's Pickup Trucks. Here Are the Top-Selling Cars of 2023," 2024, accessed February 4, 2024, https://www.cnbc.com/2024/01/06/top-10-best-selling-cars-in-the-us-in-2023.html.

1.25. EPA, "Automotive Trends Report," 2023, accessed, https://www.epa.gov/automotive-trends/download-automotive-trends-report.

1.26. Michelle Monteforte, M.R.B., Bernard, Y., Bieker, G., Lee, K. et al., "European Vehicle Market Statistics 2022/23," ICCT, 2023, accessed June 13, 2024, https://theicct.org/publication/european-vehicle-market-statistics-2022-23/.

1.27. IEA, "Fuel Economy in China," 2021, accessed February 1, 2024, https://www.iea.org/articles/fuel-economy-in-china.

1.28. Senecal, K. and Leach, F., *Racing toward Zero: The Untold Story of Driving Green* (Warrendale: SAE International, 2021).

1.29. Senecal, P.K. and Leach, F., "Diversity in Transportation: Why a Mix of Propulsion Technologies Is the Way forward for the Future Fleet," *Results in Engineering* 4 (2019): 100060, doi:https://doi.org/10.1016/j.rineng.2019.100060.

1.30. ADAC, "Elektroauto und Ladeverluste: So können Sie Kosten vermeiden," 2022, accessed February 4, 2024, https://www.adac.de/rund-ums-fahrzeug/elektromobilitaet/laden/ladeverluste-elektroauto-studie/?_x_tr_sl&_x_tr_tl&_x_tr_hl.

1.31. Strzelec, A. and Kasab, J., *Automotive Emissions Regulations and Exhaust Aftertreatment Systems* (Warrendale: SAE International, 2020).

Chapter 2

2.1. Chava, S., "Environmental Externalities and Cost of Capital," *Management Science* 60, no. 9 (2014): 2223-2247.

2.2. Baumol, W.J., "On Taxation and the Control of Externalities," *The American Economic Review* 62, no. 3 (1972): 307-322.

2.3. McMahon, W.W., "Chapter 6: The Social and External Benefits of Education," in Johnes, G. and Johnes, J. (eds.), *International Handbook on the Economics of Education* (Cheltenham, UK: Edward Elgar Publishing, 2004), vol. 211, 259.

2.4. Yang, Q., Shen, H., and Liang, Z., "Analysis of Particulate Matter and Carbon Monoxide Emission Rates from Vehicles in a Shanghai Tunnel," *Sustainable Cities and Society* 56 (2020): 102104, doi:https://doi.org/10.1016/j.scs.2020.102104.

2.5. Senecal, K. and Leach, F., *Racing toward Zero: The Untold Story of Driving Green* (Warrendale: SAE International, 2021).

2.6. Franco, V. et al., "Real-World Exhaust Emissions from Modern Diesel Cars," ICCT White Paper, October 2014, 1-59, https://theicct.org/sites/default/files/publications/ICCT_PEMS-study_diesel-cars_20141010.pdf.

2.7. Eastwood, N., "NatWest Combs Customer Accounts—And Tells Them to Go Vegetarian," *The Telegraph*, 2023, accessed February 1, 2024, https://www.telegraph.co.uk/money/banking/natwest-combs-customer-accounts-tells-them-go-vegan/.

2.8. Katanich, D., "France Bans Heated Terraces Cutting Emissions Equivalent to 300,000 Cars Every Year," euronews.green, 2021, accessed February 1, 2024, https://www.euronews.com/green/2021/11/28/why-is-this-french-city-banning-patio-heaters-in-winter.

2.9. Xanthos, D. and Walker, T.R., "International Policies to Reduce Plastic Marine Pollution from Single-Use Plastics (Plastic Bags and Microbeads): A Review," *Marine Pollution Bulletin* 118, no. 1-2 (2017): 17-26.

2.10. Mosquera, M.R., "Banning Plastic Straws: The Beginning of the War against Plastics," *Earth Jurisprudence & Envtl. Just. J.* 9 (2019): 5.

2.11. US EPA, "Global Greenhouse Gas Emissions Data," accessed February 4, 2024, https://www.epa.gov/ghgemissions/global-greenhouse-gas-emissions-data.

2.12. Xu, X. et al., "Global Greenhouse Gas Emissions from Animal-Based Foods Are Twice Those of Plant-Based Foods," *Nature Food* 2, no. 9 (2021): 724-732, doi:https://doi.org/10.1038/s43016-021-00358-x.

2.13. Menegat, S., Ledo, A., and Tirado, R., "Greenhouse Gas Emissions from Global Production and Use of Nitrogen Synthetic Fertilisers in Agriculture," *Scientific Reports* 12, no. 1 (2022): 14490, doi:https://doi.org/10.1038/s41598-022-18773-w.

2.14. Capdevila-Cortada, M., "Electrifying the Haber–Bosch," *Nature Catalysis* 2, no. 12 (2019): 1055.

2.15. Audsley, E. et al., "Estimation of the Greenhouse Gas Emissions from Agricultural Pesticide Manufacture and Use," Cranfield University, 2009.

2.16. Lundin, O. et al., "Neonicotinoid Insecticides and Their Impacts on Bees: A Systematic Review of Research Approaches and Identification of Knowledge Gaps," *PLoS One* 10, no. 8 (2015): e0136928.

2.17. Whitehorn, P.R. et al., "Neonicotinoid Pesticide Reduces Bumble Bee Colony Growth and Queen Production," *Science* 336, no. 6079 (2012): 351-352.

2.18. Capps, O. and Havlicek, J., "The Demand for Gasoline and Diesel Fuel in Agricultural Use in Virginia," *Journal of Agricultural and Applied Economics* 10, no. 1 (1978): 59-64.

2.19. Concrete Needs to Lose Its Colossal Carbon Footprint," *Nature* 597 (2021): 593-594.

2.20. Kim, J. et al., "Decarbonizing the Iron and Steel Industry: A Systematic Review of Sociotechnical Systems, Technological Innovations, and Policy Options," *Energy Research & Social Science* 89 (2022): 102565, doi:https://doi.org/10.1016/j.erss.2022.102565.

2.21. European Commission, "Certificates and Inspections," 2014, accessed February 1, 2024, https://energy.ec.europa.eu/topics/energy-efficiency/energy-efficient-buildings/certificates-and-inspections_en.

2.22. Wang, Y. and Qian, H., "Phthalates and Their Impacts on Human Health," *Healthcare* 9, no. 5 (2021): 603.

2.23. European Chemicals Agency, "Microplastics," 2023, accessed February 1, 2024, https://www.echa.europa.eu/hot-topics/microplastics.

2.24. Plastic Soup Foundation, "Recycling Codes," 2023, accessed February 1, 2024, https://www.plasticsoupfoundation.org/en/plastic-problem/what-is-plastic/recycling-codes.

2.25. MacDonald, S. and Eyre, N., "An International Review of Markets for Voluntary Green Electricity Tariffs," *Renewable and Sustainable Energy Reviews* 91 (2018): 180-192.

2.26. Burton, T., "A Data-Driven Greenhouse Gas Emission Rate Analysis for Vehicle Comparisons," *SAE Int. J. Elec. Veh.* 12, no. 1 (2023): 91-127, doi:https://doi.org/10.4271/14-12-01-0006.

2.27. IEA, "Data Centres and Data Transmission Networks," 2024, accessed February 1, 2024, https://www.iea.org/energy-system/buildings/data-centres-and-data-transmission-networks.

2.28. Marrasso, E., Roselli, C., and Sasso, M., "Electric Efficiency Indicators and Carbon Dioxide Emission Factors for Power Generation by Fossil and Renewable Energy Sources on Hourly Basis," *Energy Conversion and Management* 196 (2019): 1369-1384.

2.29. Clarke, L.B., "The Fate of Trace Elements during Coal Combustion and Gasification: An Overview," *Fuel* 72, no. 6 (1993): 731-736.

2.30. Norway, I., "Conversion Guidelines—Greenhouse Gas Emissions," accessed February 4, 2024, https://www.eeagrants.gov.pt/media/2776/conversion-guidelines.pdf.

2.31. Australia's Ozone Protection Program, *Montreal Protocol on Substances that Deplete the Ozone Layer*. Vol. 26 (Washington, DC: US Government Printing Office, 1987), 128-136.

2.32. Ritchie, H., "How the World Eliminated Lead from Gasoline," 2022, accessed February 4, 2024, https://ourworldindata.org/leaded-gasoline-phase-out.

Chapter 3

3.1. Annas, J., "Being Virtuous and Doing the Right Thing," *Proceedings and Addresses of the American Philosophical Association* 78 (2004): 61-75.

3.2. Bicchieri, C. and Xiao, E., "Do the Right Thing: But Only If Others Do So," *Journal of Behavioral Decision Making* 22, no. 2 (2009): 191-208.

3.3. Bowen, A., "'Green' Growth, 'Green' Jobs and Labor Markets," World Bank Policy Research Working Paper (5990), 2012.

3.4. Bowen, A. and Kuralbayeva, K., "Looking for Green Jobs: The Impact of Green Growth on Employment," Grantham Research Institute Working Policy Report, London School of Economics and Political Science, London, 2015, 1-28.

3.5. Lesser, J.A., "Renewable Energy and the Fallacy of 'Green' Jobs," *The Electricity Journal* 23, no. 7 (2010): 45-53, doi:https://doi.org/10.1016/j.tej.2010.06.019.

3.6.   Strietska-Ilina, O. et al., *Skills for Green Jobs: A Global View* (Geneva: International Labour Organisation, 2012).

3.7.   McCray, J.P., Gonzalez, J.J., and Darling, J.R., "Crisis Management in Smart Phones: The Case of Nokia vs Apple," *European Business Review* 23, no. 3 (2011): 240-255.

3.8.   Schumpeter, J.A., *Capitalism, Socialism and Democracy*, 3rd ed. (England, UK: Routledge, 1950).

3.9.   Harrison, M., "The Economics of World War II: An Overview," in Harrison, M. (ed.), *The Economics of World War II: Six Great Powers in International Comparison* (Cambridge, UK: Cambridge University Press, 1998), 1-42.

3.10. Xanthos, D. and Walker, T.R., "International Policies to Reduce Plastic Marine Pollution from Single-Use Plastics (Plastic Bags and Microbeads): A Review," *Marine Pollution Bulletin* 118, no. 1-2 (2017): 17-26.

3.11. Guerranti, C. et al., "Microplastics in Cosmetics: Environmental Issues and Needs for Global Bans," *Environmental Toxicology and Pharmacology* 68 (2019): 75-79.

3.12. Senecal, K. and Leach, F., *Racing toward Zero: The Untold Story of Driving Green* (Warrendale: SAE International, 2021).

3.13. Kish, R.J., "Using Legislation to Reduce One-Time Plastic Bag Usage," *Economic Affairs* 38, no. 2 (2018): 224-239.

3.14. da Cruz, N.F. et al., "Packaging Waste Recycling in Europe: Is the Industry Paying for It?" *Waste Management* 34, no. 2 (2014): 298-308.

3.15. Larcom, S., Rauch, F., and Willems, T., "The Benefits of Forced Experimentation: Striking Evidence from the London Underground Network*," *The Quarterly Journal of Economics* 132, no. 4 (2017): 2019-2055, doi:https://doi.org/10.1093/qje/qjx020.

3.16. Hinrichsen, N., "Commercially Available Alternatives to Palm Oil," *Lipid Technology* 28, no. 3-4 (2016): 65-67.

3.17. Scrinis, G. and Parker, C., "Front-of-Pack Food Labeling and the Politics of Nutritional Nudges," *Law & Policy* 38, no. 3 (2016): 234-249.

3.18. Buckton, C.H. et al., "A Discourse Network Analysis of UK Newspaper Coverage of the 'Sugar Tax' Debate before and after the Announcement of the Soft Drinks Industry Levy," *BMC Public Health* 19, no. 1 (2019): 1-14.

3.19. Haagen-Smit, A.J., "The Air Pollution Problem in Los Angeles," *Engineering and Science* 14, no. 3 (1950): 7-13.

3.20. Plumer, B., "Europe's Love Affair with Diesel Cars Has Been a Disaster," Vox, 2015, accessed February 1, 2024, https://www.vox.com/2015/10/15/9541789/volkswagen-europe-diesel-pollution.

3.21. Hooftman, N. et al., "A Review of the European Passenger Car Regulations–Real Driving Emissions vs Local Air Quality," *Renewable and Sustainable Energy Reviews* 86 (2018): 1-21.

3.22. Franco, V. et al., "Real-World Exhaust Emissions from Modern Diesel Cars," ICCT White Paper, October 2014, 1-59, https://theicct.org/sites/default/files/publications/ICCT_PEMS-study_diesel-cars_20141010.pdf.

3.23. Official Green NCAP Website, "How Green Is Your Car?," 2024, accessed February 1, 2024, https://www.greenncap.com/.

3.24. Seely, A., "Taxation of Road Fuels," House of Commons Library: 824, 2022, accessed February 1, 2024, https://researchbriefings.files.parliament.uk/documents/SN00824/SN00824.pdf.

3.25. Bolton, P., "Petrol and Diesel Prices," House of Commons Library: 4712, 2023, accessed February 1, 2024, https://researchbriefings.files.parliament.uk/documents/SN04712/SN04712.pdf.

3.26. EPA, "Automotive Trends Report," 2023, accessed June 13, 2024, https://www.epa.gov/automotive-trends/download-automotive-trends-report.

3.27. Chrystal, K.A., Mizen, P.D., and Mizen, P., "Goodhart's Law: Its Origins, Meaning and Implications for Monetary Policy," in Mizen, P. (ed.), *Central Banking, Monetary Theory and Practice: Essays in Honour of Charles Goodhart* (Cheltenham, UK: Edward Elgar Publishing, 2003), vol. 1, 221-243.

3.28. Cerruti, D., Alberini, A., and Linn, J., "Charging Drivers by the Pound: How Does the UK Vehicle Tax System Affect CO_2 Emissions?" *Environmental and Resource Economics* 74, no. 1 (2019): 99-129, doi:https://doi.org/10.1007/s10640-018-00310-x.

3.29. Raza, M. et al., "A Review of Particulate Number (PN) Emissions from Gasoline Direct Injection (GDI) Engines and Their Control Techniques," *Energies* 11, no. 6 (2018): 1417.

3.30. Kampa, M. and Castanas, E., "Human Health Effects of Air Pollution," *Environmental Pollution* 151, no. 2 (2008): 362-367.

3.31. Ma, L., Graham, D.J., and Stettler, M.E.J., "Has the Ultra Low Emission Zone in London Improved Air Quality?" *Environmental Research Letters* 16, no. 12 (2021): 124001, doi:https://doi.org/10.1088/1748-9326/ac30c1.

Chapter 4

4.1. Masala, F., "Vehicle Excise Duty," House of Commons Library: CBP-1482, 2023, accessed February 19, 2024, https://researchbriefings.files.parliament.uk/documents/SN01482/SN01482.pdf.

4.2. Unfried, A., "2022 Global Automotive Tax Guide," PwC, 2022, accessed February 27, 2024, https://www.pwc.com/ee/et/assets/document/2022-Global-Automotive_FormatierteVersion.pdf.

4.3. GOV.UK, "Fuel Duty," accessed February 19, 2024, https://www.gov.uk/tax-on-shopping/fuel-duty.

4.4. Seely, A., "Taxation of Road Fuels," House of Commons Library: 824, 2022, accessed February 1, 2024, https://researchbriefings.files.parliament.uk/documents/SN00824/SN00824.pdf.

4.5. GOV.UK, "Toll Road Charges," 2024, accessed February 27, 2024, https://www.gov.uk/uk-toll-roads.

4.6.   Antich, A.O., "Car Taxes: Europe's Powerful Tool to Accelerate Uptake of Electric Cars," 2024, accessed February 27, 2024, https://www.transportenvironment.org/car-taxes-electromobility/.

4.7.   Vries, S.d., "The Automobile Industry Pocket Guide 2023/2024," ACEA, 2023, accessed February 27, 2024, https://www.acea.auto/files/ACEA-Pocket-Guide-2023-2024.pdf.

4.8.   Kok, X., "In Singapore, a Certificate to Own a Car Now Costs $106,000," Reuters, 2023, accessed March 3, 2024, https://www.reuters.com/business/autos-transportation/singapore-certificate-own-car-now-costs-106000-2023-10-04/.

4.9.   Rosso, L. and Wagner, R.A., "How Much Does Mobility Matter for Value-Added Tax Revenue? Cross-Country Evidence around COVID-19," February 3, 2023.

4.10.   "VED Hypothecation 'Fizzles Out' after £2bn Treasury Raid," *Highways Magazine*, 2021, accessed February 27, 2024, https://www.highwaysmagazine.co.uk/VED-hypothecation-fizzles-out-after-2bn-Treasury-raid/9384.

4.11.   Office of Budget Responsibility, "Vehicle Excise Duty," 2023, accessed January 16, 2024, https://obr.uk/forecasts-in-depth/tax-by-tax-spend-by-spend/vehicle-excise-duty/.

4.12.   Office of Budget Responsibility, "Fuel Duties," 2023, accessed January 16, 2024, https://obr.uk/forecasts-in-depth/tax-by-tax-spend-by-spend/fuel-duties/.

4.13.   Senecal, K. and Leach, F., *Racing toward Zero: The Untold Story of Driving Green* (Warrendale: SAE International, 2021).

4.14.   HM Revenue and Customs, "Income Tax: Cars Appropriate Percentage—Increasing the Diesel Supplement," November 22, 2017, accessed February 27, 2024, https://www.gov.uk/government/publications/income-tax-cars-appropriate-percentage-increasing-the-diesel-supplement/income-tax-cars-appropriate-percentage-increasing-the-diesel-supplement.

4.15.   Ebrill, M.L.P., Keen, M.M., and Perry, M.V.P., *The Modern VAT* (Washington, DC: International Monetary Fund, 2001).

4.16.   Smith, A., *An Inquiry into the Nature and Causes of the Wealth of Nations*. Vol. 2 (London: Oxford University Press, 1904).

4.17.   Dresner, S., Jackson, T., and Gilbert, N., "History and Social Responses to Environmental Tax Reform in the United Kingdom," *Energy Policy* 34, no. 8 (2006): 930-939.

4.18.   Bendali, Z. et al., "Le mouvement des Gilets jaunes: un apprentissage en pratique (s) de la politique?" *Politix* 4 (2019): 143-177.

4.19.   Racu, A., "Clean and Lean: Battery Metals Demand from Electrifying Passenger Transport," Transport & Environment, 2023, accessed February 27, 2024, https://www.transportenvironment.org/wp-content/uploads/2023/07/Battery-metals-demand-from-electrifying-passenger-transport-2.pdf.

4.20.   JASIC, "Japanese Proposal and Research Plan, GRSP Inf. Group on a Pole Side Impact GTR PSI-4-11," UNECE, 2011, accessed February 28, 2024, https://unece.org/DAM/trans/main/wp29/PSI-04-11.pdf.

4.21. International Transport Forum, "1.4 Tonne Is the Average Weight of a Car," YouTube, 2021, accessed June 14, 2024, https://www.youtube.com/watch?v=PLcYYheRy60.

4.22. Parkers, "Smart Fortwo Coupe (2004–2007) Brabus 2d AutoSpecs & Dimensions," 2024, accessed February 28, 2024, https://www.parkers.co.uk/smart/fortwo/coupe-2004/brabus-2d-auto/specs/.

4.23. Huitema, E.-M., "Tax Guide," ACEA, 2022, accessed February 28, 2024, https://www.acea.auto/files/ACEA_Tax_Guide_2022.pdf.

4.24. Fleet News, "Hungary's Car Tax," 2003, accessed February 28, 2024, https://www.fleetnews.co.uk/news/2003/11/26/hungary-s-car-tax/15390/.

4.25. OECD, *Consumption Tax Trends 2020* (Paris, France: OECD Publishing, 2020).

4.26. Chen, Z., Yang, Z., and Wappelhorst, S., "Overview of Asian and Asia-Pacific Passenger Vehicle Taxation Policies and Their Potential to Drive Low-Emission Vehicle Purchases," ICCT, 2022, accessed February 28, 2024, https://theicct.org/wp-content/uploads/2022/01/Asia-Vehicle-Tax_whitepaper_final.pdf.

4.27. Ritchie, H., "The Weighty Issue of Electric Cars," 2023, accessed July 17, 2023, https://www.sustainabilitybynumbers.com/p/weighty-issue-of-electric-cars.

4.28. OECD, "Norway's Evolving Incentives for Zero-Emission Vehicles," 2022, accessed February 28, 2024, https://www.oecd.org/climate-action/ipac/practices/norway-s-evolving-incentives-for-zero-emission-vehicles-22d2485b/.

4.29. Carlier, M., "Market Share of Electric Cars (BEV and PHEV) in Norway from 2009 to 2022," September 12, 2023, accessed February 28, 2024, https://www.statista.com/statistics/1029909/market-share-of-electric-cars-in-norway/.

4.30. Carlier, M., "Market Share of Battery Electric Cars (BEV) in Norway from 2018 to 2022," September 12, 2023, accessed February 28, 2024, https://www.statista.com/statistics/1029936/market-share-of-battery-electric-cars-in-norway/.

4.31. Arba, A., "Sales Share of Electric Vehicles in Japan from 2013 to 2022," August 30, 2023, accessed February 28, 2024, https://www.statista.com/statistics/711994/japan-electric-car-market-share/.

4.32. ACEA, "New Car Registrations: +13.9% in 2023; Battery Electric 14.6% Market Share," 2024, accessed February 28, 2024, https://www.acea.auto/pc-registrations/new-car-registrations-13-9-in-2023-battery-electric-14-6-market-share/.

4.33. Slowik, P. and Isenstadt, A., "U.S. Electric Vehicle Sales Soar into '24," ICCT, 2024, accessed January 26, 2024, https://theicct.org/us-ev-sales-soar-into-24-jan24/.

4.34. Andrew, R., "Norway EV Sales and Related Data," https://robbieandrew.github.io/EV/, accessed February 28, 2024.

4.35. Bruenig, M., "5 Insights from Norway's 2023 Tax Proposal," November 21, 2022, accessed February 28, 2024, https://www.peoplespolicyproject.org/2022/11/21/5-insights-from-norways-2023-tax-proposal/.

4.36. Autodata, "2020 Hongqi E-HS9 99 kWh (551 Hp) AWD 7 Seat," accessed February 28, 2024, https://www.auto-data.net/en/hongqi-e-hs9-99-kwh-551hp-awd-7-seat-45524.

4.37. Root, V., "France to Introduce Tax on Cars Weighing over 1,800 kg: Echos," Bloomberg, 2020, accessed June 14, 2024, https://news.bloombergtax.com/daily-tax-report/france-to-introduce-tax-on-cars-weighing-over-1-800kg-echos.

4.38. Strocko, K., "French Court Rejects Challenge to Vehicle Weight Tax," *Tax Notes International* 101, no. 2 (2021): 229-230.

4.39. Liberation, "Le malus sur les voitures lourdes 'sans doute' étendu dès 2024, annonce Beaune," 2023, accessed February 28, 2024, https://www.liberation.fr/economie/transports/le-malus-sur-les-voitures-lourdes-sans-doute-etendu-des-2024-annonce-beaune-20230625_PIRNX4YXXFDW5MZUVL3KKT56TI/.

4.40. Rose, M., "Bike-Friendly Paris Votes to Triple Parking Fees for SUVs," Reuters, 2024, accessed February 28, 2024, https://www.reuters.com/sustainability/bike-friendly-paris-votes-raising-parking-fees-suvs-2024-02-03/.

4.41. State of California, Department of Motor Vehicles, "Vehicle Registration & Licensing Fee Calculators," accessed February 28, 2024, https://www.dmv.ca.gov/portal/vehicle-registration/registration-fees/vehicle-registration-fee-calculator/.

4.42. New York State, Department of Motor Vehicles, "Passenger Vehicle Registration Fees, Use Taxes and Supplemental Fees," accessed February 28, 2024, https://dmv.ny.gov/registration/registration-fees-use-taxes-and-supplemental-fees-passenger-vehicles.

4.43. Tavares, A., "Tax Cuts and Jobs Act Expands Tax Breaks for Business Vehicles," Kahn, Litwin, Renza, 2019, accessed September 16, 2019, https://kahnlitwin.com/blogs/tax-blog/tax-cuts-and-jobs-act-expands-tax-breaks-for-business-vehicles.

4.44. Land Rover Southampton, "Temporary 100% 'Bonus Depreciation' Decreases by 20% per Year Starting in 2023," accessed February 28, 2024, https://www.landroversouthampton.com/lr-tax-advantage.htm.

4.45. Toyota, "Toyota Sets out Advanced Battery Technology Roadmap," 2023, accessed February 28, 2024, https://media.toyota.co.uk/toyota-sets-out-advanced-battery-technology-roadmap/.

4.46. Williams, M. and Minjares, R., "A Technical Summary of Euro 6/VI Vehicle Emission Standards," ICCT, 2016, accessed February 1, 2024, https://theicct.org/sites/default/files/publications/ICCT_Euro6-VI_briefing_jun2016.pdf.

4.47. Gibbs, N., "ACEA Calls for Europe to Create Own KEI Car Law for Small, Urban EVs," Autocar, 2023, accessed February 28, 2024, https://www.autocar.co.uk/car-news/business-electric-vehicles/acea-calls-europe-create-own-kei-car-law-small-urban-evs.

Chapter 5

5.1. Senecal, K. and Leach, F., *Racing toward Zero: The Untold Story of Driving Green* (Warrendale: SAE International, 2021).

5.2. Williams, M. and Minjares, R., "A Technical Summary of Euro 6/VI Vehicle Emission Standards," ICCT, 2016, accessed February 1, 2024, https://theicct.org/sites/default/files/publications/ICCT_Euro6-VI_briefing_jun2016.pdf.

5.3. Uwe Tietge, S.D., Yang, Z., and Mock, P., "From Laboratory to Road International: A Comparison of Official and Real-World Fuel Consumption and CO_2 Values for Passenger Cars in Europe, the United States, China, and Japan," ICCT, 2017, accessed February 4, 2024, https://theicct.org/publication/from-laboratory-to-road-international-a-comparison-of-official-and-real-world-fuel-consumption-and-co2-values-for-passenger-cars-in-europe-the-united-states-china-and-japan/.

5.4. Emissions Analytics, "AIR Index," accessed February 1, 2024, https://airindex.com/.

5.5. Official Green NCAP Website, "How Green Is Your Car?," 2024, accessed February 1, 2024, https://www.greenncap.com/.

5.6. The Miles Consultancy, "New Analysis of Plug-In Hybrid Car MPG and Emissions Is Expected to Spark Debate on Their Suitability for Fleet Operation," 2017, accessed February 27, 2004, https://www.themilesconsultancy.com/.

5.7. Emissions Analytics, "Schrödinger's Car—Resolving the Enigma of Plug-In Hybrid Vehicles," 2020, accessed October 26, 2020, https://www.emissionsanalytics.com/news/Schr%C3%B6dinger%E2%80%99s%20Car.

5.8. Solomon, B.D., Barnes, J.R., and Halvorsen, K.E., "Grain and Cellulosic Ethanol: History, Economics, and Energy Policy," *Biomass and Bioenergy* 31, no. 6 (2007): 416-425.

5.9. Johnson, C. et al., "History of Ethanol Fuel Adoption in the United States: Policy, Economics, and Logistics," NREL/TP-5400-76260, National Renewable Energy Laboratory, 2021, accessed February 27, 2024, https://www.nrel.gov/docs/fy22osti/76260.pdf.

5.10. Shankar, V. and Leach, F., "Effects of Oxygenate and Aromatic Content on Engine-Out Aldehyde Emissions from Pure, Binary, and Ternary Mixtures of Ethanol, Toluene, and Iso-Octane," SAE Technical Paper 2023-32-0029 (2023), doi:https://doi.org/10.4271/2023-32-0029.

5.11. Shankar, V., Usen, I., Molden, N., Willman, C. et al., "Comparing Real Driving Emissions from Euro 6d-TEMP Vehicles Running on E0 and E10 Gasoline Blends," SAE Technical Paper 2023-01-1662 (2023), doi:https://doi.org/10.4271/2023-01-1662.

5.12. Ravi, S.S. et al., "On the Pursuit of Emissions-Free Clean Mobility – Electric Vehicles versus e-Fuels," *Science of the Total Environment* 875 (2023): 162688, doi:https://doi.org/10.1016/j.scitotenv.2023.162688.

5.13. Green NCAP, "Green NCAP: The Size of Your Car Does Matter," 2023, accessed February 14, 2024, https://www.greenncap.com/press-releases/green-ncap-the-size-of-your-car-does-matter/.

Chapter 6

6.1. Senecal, K. and Leach, F., *Racing toward Zero: The Untold Story of Driving Green* (Warrendale: SAE International, 2021).

6.2. EcoModder, "Vehicle Coefficient of Drag List," 2018, accessed March 5, 2024, https://ecomodder.com/wiki/Vehicle_Coefficient_of_Drag_List.

6.3. Metcalfe, H., "Pagani Huayra, the Details," 2011, accessed March 5, 2024, https://www.evo.co.uk/pagani/huayra/12129/pagani-huayra-the-details.

6.4. Tallodi, J., "10 of the Most Aerodynamic Cars Ever Made," 2024, accessed March 5, 2024, https://www.carwow.co.uk/best/most-aerodynamic-cars.

6.5. Rawlins, P., "These Are the 10 Most Aerodynamically Efficient EVs on Sale Today," accessed March 5, 2024, https://www.topgear.com/car-news/electric/these-are-10-most-aerodynamically-efficient-evs-sale-today.

6.6. Opletal, J., "Xiaomi Unveiled Its First Car Xiaomi SU7 with 800 km Range and 2.78s 0-100km/h Acceleration," 2023, accessed March 5, 2024, https://carnewschina.com/2023/12/28/xiaomi-officially-unveiled-its-first-car-xiaomi-su7-with-800-km-range-and-2-78s-0-100km-h-acceleration/.

6.7. Bickel, C.L., "Optimizing Control of Shell Eco-Marathon Prototype Vehicle to Minimize Fuel Consumption," 2017.

6.8. The Engineering ToolBox, "Rolling Resistance," accessed March 5, 2024, https://www.engineeringtoolbox.com/rolling-friction-resistance-d_1303.html.

6.9. Aldhufairi, H.S. and Olatunbosun, O.A., "Developments in Tyre Design for Lower Rolling Resistance: A State of the Art Review," *Proceedings of the Institution of Mechanical Engineers, Part D: Journal of Automobile Engineering* 232, no. 14 (2018): 1865-1882, doi:https://doi.org/10.1177/0954407017727195.

6.10. Berry, I.M., "The Effects of Driving Style and Vehicle Performance on the Real-World Fuel Consumption of US Light-Duty Vehicles," Massachusetts Institute of Technology, Cambridge, 2010.

Chapter 7

7.1. Amato, F. et al., "Urban Air Quality: The Challenge of Traffic Non-Exhaust Emissions," *Journal of Hazardous Materials* 275 (2014): 31-36.

7.2. Senecal, K. and Leach, F., *Racing toward Zero: The Untold Story of Driving Green* (Warrendale: SAE International, 2021).

7.3. IARC, *Chemical Agents and Related Occupations: A Review of Human Carcinogens* (Lyon, France: IARC, 2012).

7.4. Sarracini, F., Miyashita, E.M., and Carvalho, F.L., "NMOG Emissions Results According to PROCONVE L7 Regulation. Flex Fuel Case Study," SAE Technical Paper 2020-36-0019 (2021), doi:https://doi.org/10.4271/2020-36-0019.

7.5. Man, H. et al., "VOCs Evaporative Emissions from Vehicles in China: Species Characteristics of Different Emission Processes," *Environmental Science and Ecotechnology* 1 (2020): 100002.

7.6. La Nasa, J. et al., "Plastic Breeze: Volatile Organic Compounds (VOCs) Emitted by Degrading Macro- and Microplastics analyzed by Selected Ion Flow-Tube Mass Spectrometry," *Chemosphere* 270 (2021): 128612.

7.7. Kupiainen, K.J. and Pirjola, L., "Vehicle Non-exhaust Emissions from the Tyre-Road Interface–Effect of Stud Properties, Traction Sanding and Resuspension," *Atmospheric Environment* 45, no. 25 (2011): 4141-4146.

7.8.   Alemani, M., Wahlström, J., and Olofsson, U., "On the Influence of Car Brake System Parameters on Particulate Matter Emissions," *Wear* 396 (2018): 67-74.

7.9.   Boom, Y.J. et al., "Laboratory Evaluation of PAH and VOC Emission from Plastic-Modified Asphalt," *Journal of Cleaner Production* 377 (2022): 134489.

7.10.   Galatioto, F. et al., "Review of Road Dust Resuspension Modelling Approaches and Comparisons Analysis for a UK Case Study," *Atmosphere* 13, no. 9 (2022): 1403.

7.11.   Borbon, A. et al., "Characterisation of NMHCs in a French Urban Atmosphere: Overview of the Main Sources," *Science of the Total Environment* 292, no. 3 (2002): 177-191.

7.12.   Sepúlveda, A. et al., "A Review of the Environmental Fate and Effects of Hazardous Substances Released from Electrical and Electronic Equipments during Recycling: Examples from China and India," *Environmental Impact Assessment Review* 30, no. 1 (2010): 28-41.

7.13.   Braun, M.E. et al., "Noise Source Characteristics in the ISO 362 Vehicle Pass-By Noise Test: Literature Review," *Applied Acoustics* 74, no. 11 (2013): 1241-1265.

7.14.   Williams, M. and Minjares, R., "A Technical Summary of Euro 6/VI Vehicle Emission Standards," ICCT, 2016, accessed February 4, 2024, https://theicct.org/sites/default/files/publications/ICCT_Euro6-VI_briefing_jun2016.pdf.

7.15.   Mock, P. et al., "The WLTP: How a New Test Procedure for Cars Will Affect Fuel Consumption Values in the EU," *International Council on Clean Transportation* 9, no. 3547 (2014): 1-20.

7.16.   Degraeuwe, B. and Weiss, M., "Does the New European Driving Cycle (NEDC) Really Fail to Capture the NOX Emissions of Diesel Cars in Europe?" *Environmental Pollution* 222 (2017): 234-241.

7.17.   Marotta, A. et al., "Gaseous Emissions from Light-Duty Vehicles: Moving from NEDC to the New WLTP Test Procedure," *Environmental Science & Technology* 49, no. 14 (2015): 8315-8322.

7.18.   Dornoff, J., Valverde Morales, V., Tietge U., "On the Way to 'Real-World' CO_2 Values? The European Passenger Car Market after 5 Years of WLTP," ICCT White Paper, 2024, accessed June 24, 2024, https://theicct.org/publication/real-world-co2-emission-values-vehicles-europe-jan24/.

7.19.   Tietge, U., Díaz, S., Yang, Z., and Mock, P., "From Laboratory to Road International: A Comparison of Official and Real-World Fuel Consumption and CO_2 Values for Passenger Cars in Europe, the United States, China, and Japan," ICCT, 2017, accessed February 4, 2024, https://theicct.org/publication/from-laboratory-to-road-international-a-comparison-of-official-and-real-world-fuel-consumption-and-co2-values-for-passenger-cars-in-europe-the-united-states-china-and-japan/.

7.20.   AIR, "The AIR Index," accessed February 1, 2024, https://airindex.com/.

7.21.   Haagen-Smit, A.J., "The Air Pollution Problem in Los Angeles," *Engineering and Science* 14, no. 3 (1950): 7-13.

7.22.   Sircar, K. et al., "Carbon Monoxide Poisoning Deaths in the United States, 1999 to 2012," *The American Journal of Emergency Medicine* 33, no. 9 (2015): 1140-1145.

7.23. Stone, R., *Introduction to Internal Combustion Engines*. 3rd Ed. (London, UK: Springer, 1999).

7.24. Heywood, J.B., "Combustion Engine Fundamentals. 1ª Edição," *Estados Unidos* 25 (1988): 1117-1128.

7.25. Harrison, R.M., *Handbook of Air Pollution Analysis* (Dordrecht: Springer Science & Business Media, 2012).

7.26. Calle-Asensio, A. et al., "Effect of Advanced Biofuels on WLTC Emissions of a Euro 6 Diesel Vehicle with SCR under Different Climatic Conditions," *International Journal of Engine Research* 22, no. 12 (2021): 3433-3446.

7.27. Raza, M. et al., "A Review of Particulate Number (PN) Emissions from Gasoline Direct Injection (GDI) Engines and Their Control Techniques," *Energies* 11, no. 6 (2018): 1417.

7.28. Guan, B. et al., "Review of the State-of-the-Art of Exhaust Particulate Filter Technology in Internal Combustion Engines," *Journal of Environmental Management* 154 (2015): 225-258.

7.29. Khair, M., "A Review of Diesel Particulate Filter Technologies," SAE Technical Paper 2003-01-2303 (2003), doi:https://doi.org/10.4271/2003-01-2303.

7.30. Joshi, A. and Johnson, T.V., "Gasoline Particulate Filters—A Review," *Emission Control Science and Technology* 4, no. 4 (2018): 219-239.

7.31. Joshi, A., "Review of Vehicle Engine Efficiency and Emissions," *SAE Int. J. Adv. & Curr. Prac. in Mobility* 2, no. 5 (2020): 2479-2507, doi:https://doi.org/10.4271/2020-01-0352.

7.32. Ravishankara, A.R., Daniel, J.S., and Portmann, R.W., "Nitrous Oxide (N_2O): The Dominant Ozone-Depleting Substance Emitted in the 21st Century," *Science* 326, no. 5949 (2009): 123-125, doi:https://doi.org/10.1126/science.1176985.

7.33. Shankar, V. and Leach, F., "Effects of Oxygenate and Aromatic Content on Engine-Out Aldehyde Emissions from Pure, Binary, and Ternary Mixtures of Ethanol, Toluene, and Iso-octane," SAE Technical Paper 2023-32-0029 (2023), doi:https://doi.org/10.4271/2023-32-0029.

7.34. Ritchie, H., "The Weighty Issue of Electric Cars," 2023, accessed July 17, 2023, https://www.sustainabilitybynumbers.com/p/weighty-issue-of-electric-cars.

7.35. Senecal, P.K. and Leach, F., "Diversity in Transportation: Why a Mix of Propulsion Technologies Is the Way forward for the Future Fleet," *Results in Engineering* 4 (2019): 100060, doi:https://doi.org/10.1016/j.rineng.2019.100060.

7.36. Ghaffarpasand, O. et al., "Real-World Assessment of Vehicle Air Pollutant Emissions Subset by Vehicle Type, Fuel and EURO Class: New Findings from the Recent UK EDAR Field Campaigns, and Implications for Emissions Restricted Zones," *Science of the Total Environment* 734 (2020): 139416, doi:https://doi.org/10.1016/j.scitotenv.2020.139416.

Chapter 8

8.1. Read, L.E. and Friedman, M., *I, Pencil: My Family Tree* (New York: Foundation for Economic Education, 1958).

8.2. Del Pero, F., Delogu, M., and Pierini, M., "Life Cycle Assessment in the Automotive Sector: A Comparative Case Study of Internal Combustion Engine (ICE) and Electric Car," *Procedia Structural Integrity* 12 (2018): 521-537.

8.3. Jones, P. and Sapsford, S., "Accelerating Road Transport Decarbonisation," Institution of Mechanical Engineers, 2020, accessed February 14, 2024, https://www.imeche.org/policy-and-press/reports/detail/accelerating-road-transport-decarbonisation.

8.4. Morris, E., "From Horse Power to Horsepower," *Access Magazine* 1, no. 30 (2007): 2-10.

8.5. Senecal, K. and Leach, F., *Racing toward Zero: The Untold Story of Driving Green* (Warrendale: SAE International, 2021).

8.6. Hawkins, T.R. et al., "Comparative Environmental Life Cycle Assessment of Conventional and Electric Vehicles," *Journal of Industrial Ecology* 17, no. 1 (2013): 53-64.

8.7. Prieur, A., Bouvart, F., Gabrielle, B., and Lehuger, S., "Well to Wheels Analysis of Biofuels vs. Conventional Fossil Fuels: A Proposal for Greenhouse Gases and Energy Savings Accounting in the French Context," SAE Technical Paper 2008-01-0673 (2008), doi:https://doi.org/10.4271/2008-01-0673.

8.8. Ramachandran, S. and Stimming, U., "Well to Wheel Analysis of Low Carbon Alternatives for Road Traffic," *Energy & Environmental Science* 8, no. 11 (2015): 3313-3324.

8.9. Yoo, E., Kim, M., and Song, H.H., "Well-to-Wheel Analysis of Hydrogen Fuel-Cell Electric Vehicle in Korea," *International Journal of Hydrogen Energy* 43, no. 41 (2018): 19267-19278.

8.10. Prussi, M. et al., "Comparing e-Fuels and Electrification for Decarbonization of Heavy-Duty Transports," *Energies* 15, no. 21 (2022): 8075.

8.11. Ueckerdt, F. et al., "Potential and Risks of Hydrogen-Based e-Fuels in Climate Change Mitigation," *Nature Climate Change* 11, no. 5 (2021): 384-393.

8.12. Grunditz, E.A. and Thiringer, T., "Performance Analysis of Current BEVs Based on a Comprehensive Review of Specifications," *IEEE Transactions on Transportation Electrification* 2, no. 3 (2016): 270-289.

8.13. Burton, T., Powers, S., Burns, C., Conway, G. et al., "A Data-Driven Greenhouse Gas Emission Rate Analysis for Vehicle Comparisons," *SAE Int. J. Elec. Veh.* 12, no. 1 (2023): 91-127, doi:https://doi.org/10.4271/14-12-01-0006.

8.14. Warner, E. et al., "Challenges in the Estimation of Greenhouse Gas Emissions from Biofuel-Induced Global Land-Use Change," *Biofuels, Bioproducts and Biorefining* 8, no. 1 (2014): 114-125.

8.15. Klöpffer, W., "Life Cycle Assessment: From the Beginning to the Current State," *Environmental Science and Pollution Research* 4 (1997): 223-228.

8.16. Pryshlakivsky, J. and Searcy, C., "Fifteen Years of ISO 14040: A Review," *Journal of Cleaner Production* 57 (2013): 115-123.

8.17. Concawe, "The Cars CO_2 Comparator," accessed February 14, 2024, https://www.carsco2comparator.eu/.

8.18. Argonne National Laboratory, "The Greenhouse Gases, Regulated Emissions, and Energy Use in Technologies Model," accessed February 14, 2024, https://greet.anl.gov/.

8.19. White House, "Treasury Sets out Proposed Rules for Transformative Clean Hydrogen Incentives," 2023, accessed February 14, 2024, https://www.whitehouse.gov/cleanenergy/clean-energy-updates/2023/12/22/treasury-sets-out-proposed-rules-for-transformative-clean-hydrogen-incentives/.

8.20. California Air Resources Board, "LCFS Life Cycle Analysis Models and Documentation," accessed February 14, 2024, https://ww2.arb.ca.gov/resources/documents/lcfs-life-cycle-analysis-models-and-documentation.

8.21. Jungmeier, D.G., Canella, L., and Schwarzinger, S., "Estimated Greenhouse Gas Emissions and Primary Energy Demand of Passenger Vehicles," Green NCAP, 2023, accessed February 14, 2024, https://www.greenncap.com/wp-content/uploads/Green-NCAP-Life-Cycle-Assessment-Methodology-and-Data_2nd-edition.pdf.

8.22. Helmers, E. and Weiss, M., "Advances and Critical Aspects in the Life-Cycle Assessment of Battery Electric Cars," *Energy and Emission Control Technologies* 5 (2017): 1-18.

8.23. Daimler, "New Vehicle Generation Provides Foundation for Growth—2015: Smart Accelerates Its Success," 2015, accessed February 14, 2024, https://web.archive.org/web/20150703053726/http:/media.daimler.com/dcmedia/0-921-614239-1-1778066-1-0-0-0-0-0-0-0-0-0-0-0-0-0-0.html.

8.24. Wikipedia, "Ford Focus: Sales," 2024, accessed February 14, 2024, https://en.wikipedia.org/wiki/Ford_Focus#Sales.

8.25. Taub, A.I. et al., "The Evolution of Technology for Materials Processing over the last 50 Years: The Automotive Example," *JOM* 59, no. 2 (2007): 48-57, doi:https://doi.org/10.1007/s11837-007-0022-7.

8.26. Luk, J.M. et al., "Greenhouse Gas Emission Benefits of Vehicle Lightweighting: Monte Carlo Probabalistic Analysis of the Multi Material Lightweight Vehicle Glider," *Transportation Research Part D: Transport and Environment* 62 (2018): 1-10, doi:https://doi.org/10.1016/j.trd.2018.02.006.

8.27. Just Auto, "Carbon Fibre in Car Production—Weighing up the Benefits," *Just Auto Magazine*, 2020, accessed February 14, 2024, https://justauto.nridigital.com/just-auto_magazine_jun20/carbon_fibre_in_car_production_weighing_up_the_benefits.

8.28. Lukaszewicz, D., "Design Drivers for Enhanced Crash Performance of Automotive CFRP Structures," in *23rd International Technical Conference on the Enhanced Safety of Vehicles*, Seoul, South Korea, 2013.

8.29. Scheyrer, J., "How Much CO_2 Is Saved When an Aluminum Automobile Frame Is Used Instead of a Steel One?," Sustamize, 2023, accessed December 7, 2023, https://www.sustamize.com/blog/how-much-co2-is-saved-when-an-aluminum-automobile-frame-is-used-instead-of-a-steel-one.

8.30. Sustainable Ships, "What Is the Carbon Footprint of Steel?," accessed February 14, 2024, https://www.sustainable-ships.org/stories/2022/carbon-footprint-steel#:~:text=The%20 IEA%20estimates%20that%20direct,of%20CO2%20per%20ton%20steel.

8.31. alupro, "Carbon Footprint of Aluminium," 2018, accessed February 14, 2024, https:// alupro.org.uk/sustainability/fact-sheets/carbon-footprint/.

8.32. Kawajiri, K. and Sakamoto, K., "Environmental Impact of Carbon Fibers Fabricated by an Innovative Manufacturing Process on Life Cycle Greenhouse Gas Emissions," *Sustainable Materials and Technologies* 31 (2022): e00365, doi:https://doi.org/10.1016/j.susmat.2021.e00365.

8.33. Al Hawari, A. et al., "A Life Cycle Assessment (LCA) of Aluminum Production Process," *International Journal of Environmental and Ecological Engineering* 8, no. 4 (2014): 704-710.

8.34. Van Caneghem, J. et al., "Improving Eco-Efficiency in the Steel Industry: The ArcelorMittal Gent Case," *Journal of Cleaner Production* 18, no. 8 (2010): 807-814, doi:https://doi.org/10.1016/j.jclepro.2009.12.016.

8.35. Cheng, H. et al., "A Closed-Loop Recycling Process for Carbon Fiber-Reinforced Polymer Waste Using Thermally Activated Oxide Semiconductors: Carbon Fiber Recycling, Characterization and Life Cycle Assessment," *Waste Management* 153 (2022): 283-292.

8.36. Mekhilef, S., Saidur, R., and Safari, A., "Comparative Study of Different Fuel Cell Technologies," *Renewable and Sustainable Energy Reviews* 16, no. 1 (2012): 981-989.

8.37. Boretti, A., "Hydrogen Internal Combustion Engines to 2030," *International Journal of Hydrogen Energy* 45, no. 43 (2020): 23692-23703.

8.38. Verhelst, S. and Wallner, T., "Hydrogen-Fueled Internal Combustion Engines," *Progress in Energy and Combustion Science* 35, no. 6 (2009): 490-527.

8.39. Boretti, A., "Progress in Cold/Cryo-Pressurized Composite Tanks for Hydrogen," *MRS Communications* 13 (2023): 400-405.

8.40. Ntombela, M., Musasa, K., and Moloi, K., "A Comprehensive Review for Battery Electric Vehicles (BEV) Drive Circuits Technology, Operations, and Challenges," *World Electric Vehicle Journal* 14, no. 7 (2023): 195.

8.41. Duan, J. et al., "Building Safe Lithium-Ion Batteries for Electric Vehicles: A Review," *Electrochemical Energy Reviews* 3 (2020): 1-42.

8.42. Kane, M., "Check Electric Cars Listed by Weight per Battery Capacity (kWh)," Inside EVs, 2021, accessed August 23, 2021, https://insideevs.com/news/528346/ev-weight-per-battery-capacity/.

8.43. Crawford, I., "How Much CO_2 Is Emitted by Manufacturing Batteries?," MIT Climate Portal, 2022, accessed July 15, 2022, https://climate.mit.edu/ask-mit/how-much-co2-emitted-manufacturing-batteries.

8.44. Aichberger, C. and Jungmeier, G., "Environmental Life Cycle Impacts of Automotive Batteries Based on a Literature Review," *Energies* 13, no. 23 (2020): 6345.

8.45. Sullivan, J., Kelly, J., and Elgowainy, A., "Vehicle Materials: Material Composition of Powertrain Systems," Argonne National Laboratory, Lemont, IL, 2018, accessed February 14, 2024, https://greet.anl.gov/files/2015-powertrain-materials.

8.46. Mruzek, M. et al., "Analysis of Parameters Influencing Electric Vehicle Range," *Procedia Engineering* 134 (2016): 165-174, doi:https://doi.org/10.1016/j.proeng.2016.01.056.

8.47. Hong, J. et al., "Life Cycle Assessment of Copper Production: A Case Study in China," *The International Journal of Life Cycle Assessment* 23 (2018): 1814-1824.

8.48. Lombardi, L. et al., "Comparative Environmental Assessment of Conventional, Electric, Hybrid, and Fuel Cell Powertrains Based on LCA," *The International Journal of Life Cycle Assessment* 22 (2017): 1989-2006.

8.49. Kim, H.C. et al., "Life Cycle Assessment of Vehicle Lightweighting: Novel Mathematical Methods to Estimate Use-Phase Fuel Consumption," *Environmental Science & Technology* 49, no. 16 (2015): 10209-10216, doi:https://doi.org/10.1021/acs.est.5b01655.

8.50. Sullivan, J.L. and Cobas-Flores, E., "Full Vehicle LCAs: A Review," SAE Technical Paper 2001-01-3725 (2001), doi:https://doi.org/10.4271/2001-01-3725.

8.51. Kim, H.C. and Wallington, T.J., "Life Cycle Assessment of Vehicle Lightweighting: A Physics-Based Model of Mass-Induced Fuel Consumption," *Environmental Science & Technology* 47, no. 24 (2013): 14358-14366, doi:https://doi.org/10.1021/es402954w.

8.52. Hottle, T. et al., "Critical Factors Affecting Life Cycle Assessments of Material Choice for Vehicle Mass Reduction," *Transportation Research Part D: Transport and Environment* 56 (2017): 241-257, doi:https://doi.org/10.1016/j.trd.2017.08.010.

8.53. Suzuki, T. and Takahashi, J., "LCA of Lightweight Vehicles by Using CFRP for Mass-Produced Vehicles," in *Fifteenth International Conference on Composite Materials*, Durban, South Africa, 2005.

8.54. Del Pero, F. et al., "Automotive Lightweight Design: Simulation Modeling of Mass-Related Consumption for Electric Vehicles," *Machines* 8, no. 3 (2020): 51.

8.55. Green NCAP, "Green NCAP: The Size of Your Car Does Matter," 2023, accessed February 14, 2024, https://www.greenncap.com/press-releases/green-ncap-the-size-of-your-car-does-matter/.

Chapter 9

9.1. Senecal, K. and Leach, F., *Racing toward Zero: The Untold Story of Driving Green* (Warrendale: SAE International, 2021).

9.2. Grant, A., "WLTP Unfairly Advantaging Petrol Engines, Study Shows," Fleet World, 2019, accessed February 9, 2024, https://fleetworld.co.uk/wltp-unfairly-advantaging-petrol-engines-study-shows/.

9.3. Harrison, R.M., *Handbook of Air Pollution Analysis* (Dordrecht: Springer Science & Business Media, 2012).

9.4. Tian, Z. et al., "A Ubiquitous Tire Rubber–Derived Chemical Induces Acute Mortality in Coho Salmon," *Science* 371, no. 6525 (2021): 185-189, doi:https://doi.org/10.1126/science.abd6951.

9.5.   Castan, S. et al., "Uptake, Metabolism, and Accumulation of Tire Wear Particle-Derived Compounds in Lettuce," *Environmental Science & Technology* 57, no. 1 (2023): 168-178, doi:https://doi.org/10.1021/acs.est.2c05660.

9.6.   Du, B. et al., "First Report on the Occurrence of N-(1,3-Dimethylbutyl)-N′-phenyl-p-phenylenediamine (6PPD) and 6PPD-Quinone as Pervasive Pollutants in Human Urine from South China," *Environmental Science & Technology Letters* 9, no. 12 (2022): 1056-1062, doi:https://doi.org/10.1021/acs.estlett.2c00821.

9.7.   Raza, M. et al., "A Review of Particulate Number (PN) Emissions from Gasoline Direct Injection (GDI) Engines and Their Control Techniques," *Energies* 11, no. 6 (2018): 1417.

9.8.   Guan, B. et al., "Review of the State-of-the-Art of Exhaust Particulate Filter Technology in Internal Combustion Engines," *Journal of Environmental Management* 154 (2015): 225-258.

9.9.   ICCT, "What EPA's New Multi-Pollutant Emissions Proposal Means for PM Emissions and GPFs," 2023, accessed February 9, 2024, https://theicct.org/wp-content/uploads/2023/11/ID-48-%E2%80%93-U.S.-GPF-fact-sheet-letter-70112-v3.pdf.

9.10. Giechaskiel, B. et al., "Effect of Tampering on On-Road and Off-Road Diesel Vehicle Emissions," *Sustainability* 14, no. 10 (2022): 6065.

9.11. Barrie, L.A., "Arctic Air Pollution: An Overview of Current Knowledge," *Atmospheric Environment (1967)* 20, no. 4 (1986): 643-663, doi:https://doi.org/10.1016/0004-6981(86)90180-0.

9.12. De Gennaro, M., Paffumi, E., and Martini, G., "Data-Driven Analysis of the Effectiveness of Evaporative Emissions Control Systems of Passenger Cars in Real World Use Condition: Time and Spatial Mapping," *Atmospheric Environment* 129 (2016): 277-293, doi:https://doi.org/10.1016/j.atmosenv.2016.01.026.

9.13. Lu, Q., Zhao, Y., and Robinson, A.L., "Comprehensive Organic Emission Profiles for Gasoline, Diesel, and Gas-Turbine Engines Including Intermediate and Semi-Volatile Organic Compound Emissions," *Atmospheric Chemistry and Physics* 18, no. 23 (2018): 17637-17654.

9.14. Brown, R., "Compound Types in Gasoline by Mass Spectrometer Analysis," *Analytical Chemistry* 23, no. 3 (1951): 430-437.

9.15. California Air Resources Board, "California Evaporative Emission Standards and Test Procedures for 1978-2000 Model Motor Vehicles," 1999, accessed February 9, 2024, https://ww2.arb.ca.gov/sites/default/files/2019-08/evaptp%201978-2000%20MY%20%28amended%208-5-1999%29_accessible.pdf.

9.16. Hawley, D., "What Is An Evaporative Control System?," JD Power, 2023, accessed February 9, 2024, https://www.jdpower.com/cars/shopping-guides/what-is-an-evaporative-control-system.

9.17. Bureau of Transportation Statistics, "Estimated U.S. Average Vehicle Emissions Rates per Vehicle by Vehicle Type Using Gasoline and Diesel," 2023, accessed February 9, 2024, https://www.bts.gov/content/estimated-national-average-vehicle-emissions-rates-vehicle-vehicle-type-using-gasoline-and.

9.18.   ECCC, "Common Air Pollutants: Volatile Organic Compounds," 2013, accessed February 9, 2024, https://www.canada.ca/en/environment-climate-change/services/air-pollution/pollutants/common-contaminants/volatile-organic-compounds.html.

9.19.   Rienda, I.C. and Alves, C.A., "Road Dust Resuspension: A Review," *Atmospheric Research* 261 (2021): 105740.

9.20.   Duncan, S. et al., "Transport of Road-Surface Sediment through Ephemeral Stream Channels 1," *JAWRA Journal of the American Water Resources Association* 23, no. 1 (1987): 113-119.

9.21.   Schwab, J.A. and Zenkel, M., "Filtration of Particulates in the Human Nose," *The Laryngoscope* 108, no. 1 (1998): 120-124.

9.22.   Ristovski, Z.D. et al., "Respiratory Health Effects of Diesel Particulate Matter," *Respirology* 17, no. 2 (2012): 201-212.

9.23.   Sehmel, G., "Particle Resuspension from an Asphalt Road Caused by Car and Truck Traffic," *Atmospheric Environment (1967)* 7, no. 3 (1973): 291-309.

9.24.   Thorpe, A.J. et al., "Estimation of Particle Resuspension Source Strength on a Major London Road," *Atmospheric Environment* 41, no. 37 (2007): 8007-8020.

9.25.   Amato, F. et al., "Traffic Induced Particle Resuspension in Paris: Emission Factors and Source Contributions," *Atmospheric Environment* 129 (2016): 114-124.

9.26.   Barlow, T. et al., "Non-Exhaust Particulate Matter Emissions from Road Traffic: Summary Report," 2007, accessed February 9, 2024, https://uk-air.defra.gov.uk/assets/documents/reports/cat15/0706061644_Report5_Summary.pdf.

9.27.   Grigoratos, T. and Martini, G., "Brake Wear Particle Emissions: A Review," *Environ Sci Pollut Res Int* 22, no. 4 (2015): 2491-2504, doi:https://doi.org/10.1007/s11356-014-3696-8.

9.28.   Piscitello, A. et al., "Non-exhaust Traffic Emissions: Sources, Characterization, and Mitigation Measures," *Science of the Total Environment* 766 (2021): 144440, doi:https://doi.org/10.1016/j.scitotenv.2020.144440.

9.29.   Garg, B.D. et al., "Brake Wear Particulate Matter Emissions," *Environmental Science & Technology* 34, no. 21 (2000): 4463-4469, doi:https://doi.org/10.1021/es001108h.

9.30.   Bondorf, L. et al., "Airborne Brake Wear Emissions from a Battery Electric Vehicle," *Atmosphere* 14, no. 3 (2023): 488.

9.31.   Gao, Y., Chen, L., and Ehsani, M., "Investigation of the Effectiveness of Regenerative Braking for EV and HEV," *SAE Transactions* 108 (1999): 3184-3190.

9.32.   California Air Resources Board, "Brake & Tire Wear Emissions," accessed February 13, 2024, https://ww2.arb.ca.gov/resources/documents/brake-tire-wear-emissions.

9.33.   Zhang, M. et al., "A Comprehensive Review of Tyre Wear Particles: Formation, Measurements, Properties, and Influencing Factors," *Atmospheric Environment* 297 (2023): 119597, doi:https://doi.org/10.1016/j.atmosenv.2023.119597.

9.34.   Kreider, M.L. et al., "Physical and Chemical Characterization of Tire-Related Particles: Comparison of Particles Generated Using Different Methodologies," *Science of the Total Environment* 408, no. 3 (2010): 652-659, doi:https://doi.org/10.1016/j.scitotenv.2009.10.016.

9.35.   Knight, L.J. et al., "Tyre Wear Particles: An Abundant Yet Widely Unreported Microplastic?" *Environmental Science and Pollution Research* 27, no. 15 (2020): 18345-18354, doi:https://doi.org/10.1007/s11356-020-08187-4.

9.36.   Emissions Analytics, "Side Effects May Include…," 2020, accessed December 3, 2020, https://www.emissionsanalytics.com/news/side-effects-may-include.

9.37.   Emissions Analytics, "Do No Harm," 2023, accessed March 20, 2023, https://www.emissionsanalytics.com/news/do-no-harm.

9.38.   Tagesson, K., Jacobson, B., and Laine, L., "Driver Response at Tyre Blow-Out in Heavy Vehicles & the Importance of Scrub Radius," in *2014 IEEE Intelligent Vehicles Symposium Proceedings*, Dearborn, MI, 2014, IEEE.

9.39.   Le Maitre, O., Süssner, M., and Zarak, C., "Evaluation of Tire Wear Performance," SAE Technical Paper 980256 (1998), doi:https://doi.org/10.4271/980256.

9.40.   Liu, Y. et al., "Impact of Vehicle Type, Tyre Feature and Driving Behaviour on Tyre Wear under Real-World Driving Conditions," *Science of the Total Environment* 842 (2022): 156950.

9.41.   Baensch-Baltruschat, B. et al., "Tyre and Road Wear Particles (TRWP)—A Review of Generation, Properties, Emissions, Human Health Risk, Ecotoxicity, and Fate in the Environment," *Science of the Total Environment* 733 (2020): 137823.

9.42.   Alves, C.A. et al., "Physical and Chemical Properties of Non-exhaust Particles Generated from Wear between Pavements and Tyres," *Atmospheric Environment* 224 (2020): 117252, doi:https://doi.org/10.1016/j.atmosenv.2019.117252.

9.43.   European Tyre and Rubber Manufacturers' Association, "Addressing Tyre and Road Wear Particles," accessed February 13, 2024, https://www.tyreandroadwear.com/.

9.44.   Zimmer, R., Reeser, W., and Cummins, P., "Evaluation of PM10 Emission Factors for Paved Streets," PM10 Standards and Nontraditional Particulate Source Controls Volume I, 1992.

9.45.   Gustafsson, M. et al., "Factors Influencing PM10 Emissions from Road Pavement Wear," *Atmospheric Environment* 43, no. 31 (2009): 4699-4702, doi:https://doi.org/10.1016/j.atmosenv.2008.04.028.

9.46.   Williams, M. and Minjares, R., "A Technical Summary of Euro 6/VI Vehicle Emission Standards," ICCT, 2016, accessed February 1, 2024, https://theicct.org/sites/default/files/publications/ICCT_Euro6-VI_briefing_jun2016.pdf.

9.47.   Emissions Analytics, "Gaining Traction, Losing Tread Pollution from Tire Wear Now 1,850 Times Worse than Exhaust Emissions," 2022, accessed May 10, 2022, https://www.emissionsanalytics.com/news/gaining-traction-losing-tread.

9.48.   Lewis, A., Moller, S.J., and Carslaw, D., "Non-exhaust Emissions from Road Traffic," 2019.

9.49.   Eunomia, "Plastics in the Marine Environment," 2019, accessed February 12, 2024, https://eunomia.eco/reports/plastics-in-the-marine-environment/.

Chapter 10

10.1. Halonen, J.I. et al., "Road Traffic Noise Is Associated with Increased Cardiovascular Morbidity and Mortality and All-Cause Mortality in London," *European Heart Journal* 36, no. 39 (2015): 2653-2661, doi:https://doi.org/10.1093/eurheartj/ehv216.

10.2. Smith, R.B. et al., "Impact of London's Road Traffic Air and Noise Pollution on Birth Weight: Retrospective Population Based Cohort Study," *BMJ* 359 (2017): j5299, doi:https://doi.org/10.1136/bmj.j5299.

10.3. Cai, Y. et al., "Impact of Road Traffic Noise on Obesity Measures: Observational Study of Three European Cohorts," *Environmental Research* 191 (2020): 110013, doi:https://doi.org/10.1016/j.envres.2020.110013.

10.4. Halperin, D., "Environmental Noise and Sleep Disturbances: A Threat to Health?" *Sleep Science* 7, no. 4 (2014): 209-212, doi:https://doi.org/10.1016/j.slsci.2014.11.003.

10.5. European Environmental Agency, "Environmental Noise in Europe—2020," EEA Report No. 22/2019, 2020.

10.6. Elliott, H., "Rolls-Royce's New Car Was So Quiet at First, It Nauseated Drivers," Bloomberg, 2020, accessed February 12, 2024, https://www.bloomberg.com/news/articles/2020-10-16/rolls-royce-s-new-car-was-so-quiet-at-first-it-made-drivers-nauseous?leadSource=uverify%20wall.

10.7. Olson, N., "Survey of Motor Vehicle Noise," *The Journal of the Acoustical Society of America* 52, no. 5A (2005): 1291-1306, doi:https://doi.org/10.1121/1.1913246.

10.8. Waters, P.E., "Commercial Road Vehicle Noise," *Journal of Sound and Vibration* 35, no. 2 (1974): 155-222, doi:https://doi.org/10.1016/0022-460X(74)90047-9.

10.9. Peng, J. et al., "Influence of Translational Vehicle Dynamics on Heavy Vehicle Noise Emission," *Science of the Total Environment* 689 (2019): 1358-1369, doi:https://doi.org/10.1016/j.scitotenv.2019.06.426.

10.10. Knowles, D., "Carmageddon," 2023.

10.11. Klein, T.m., Hertz, E., and Borener, S., "A Collection of Recent Analyses of Vehicle Weight and Safety," in *Proceedings: International Technical Conference on the Enhanced Safety of Vehicles*, Yokohama, Japan, 1993, 94-103.

10.12. Høye, A., "Vehicle Registration Year, Age, and Weight—Untangling the Effects on Crash Risk," *Accident Analysis & Prevention* 123 (2019): 1-11, doi:https://doi.org/10.1016/j.aap.2018.11.002.

10.13. Wenzel, T. and Ross, M., "The Relationship between Vehicle Weight/Size and Safety," *AIP Conference Proceedings* 1044, no. 1 (2008): 251-265, doi:https://doi.org/10.1063/1.2993724.

10.14. Mahdinia, I., Khattak, A.J., and Mohsena Haque, A., "How Effective Are Pedestrian Crash Prevention Systems in Improving Pedestrian Safety? Harnessing Large-Scale Experimental Data," *Accident Analysis & Prevention* 171 (2022): 106669, doi:https://doi.org/10.1016/j.aap.2022.106669.

10.15. Young, R.D., Ross, H., and Lammert, W.F., "Simulation of the Pedestrian during Vehicle Impact," in *Proceedings of the 3rd International Congress on Automotive Safety*, San Francisco, 1974.

10.16. Simms, C.K. and Wood, D.P., "Pedestrian Risk from Cars and Sport Utility Vehicles—A Comparative Analytical Study," *Proceedings of the Institution of Mechanical Engineers, Part D: Journal of Automobile Engineering* 220, no. 8 (2006): 1085-1100, doi:https://doi.org/10.1243/09544070JAUTO319.

10.17. Tyndall, J., "Pedestrian Deaths and Large Vehicles," *Economics of Transportation* 26-27 (2021): 100219, doi:https://doi.org/10.1016/j.ecotra.2021.100219.

10.18. Tyndall, J., "The Effect of Front-End Vehicle Height on Pedestrian Death Risk," *Economics of Transportation* 37 (2024): 100342, doi:https://doi.org/10.1016/j.ecotra.2024.100342.

10.19. Edwards, M. and Leonard, D., "Effects of Large Vehicles on Pedestrian and Pedalcyclist Injury Severity," *Journal of Safety Research* 82 (2022): 275-282, doi:https://doi.org/10.1016/j.jsr.2022.06.005.

10.20. Evans, L. and Frick, M.C., "Car Size or Car Mass: Which Has Greater Influence on Fatality Risk?" *Am J Public Health* 82, no. 8 (1992): 1105-1112, doi:https://doi.org/10.2105/ajph.82.8.1105.

10.21. Evans, L. and Frick, M.C., "Mass Ratio and Relative Driver Fatality Risk in Two-Vehicle Crashes," *Accident Analysis & Prevention* 25, no. 2 (1993): 213-224, doi:https://doi.org/10.1016/0001-4575(93)90062-2.

10.22. Padmanaban, J., "Influences of Vehicle Size and Mass and Selected Driver Factors on Odds of Driver Fatality," *Annu Proc Assoc Adv Automot Med* 47 (2003): 507-524.

10.23. Moore, T.C. and Lovins, A.B., "Vehicle Design Strategies to Meet and Exceed PNGV Goals," *SAE Transactions* 104 (1995): 2676-2718.

10.24. Cebon, D., "Road Damaging Effects of Dynamic Axle Loads," in *Proceedings, International Symposium on Heavy Vehicle Weights and Dimensions*, Kelowna, British Columbia, 1986.

10.25. Rhodes, A.H. and Suara, A.M., "The Structural Wear of Road Pavements: An Assessment of the Fourth Power Law on the A1(M) Motorway in County Durham," *Proceedings of the Institution of Civil Engineers – Transport* 105, no. 4 (1994): 273-281, doi:https://doi.org/10.1680/itran.1994.27138.

10.26. Cebon, D., "Vehicle-Generated Road Damage: A Review," *Vehicle System Dynamics* 18, no. 1-3 (1989): 107-150, doi:https://doi.org/10.1080/00423118908968916.

10.27. Hope, C. and Simpson, J., "Sheer Weight of Electric Vehicles Could Sink Our Bridges," *The Telegraph*, 2023, accessed February 12, 2024, https://www.telegraph.co.uk/news/2023/05/07/electric-vehicles-33-per-cent-heavier-bridges-collapse/.

10.28. HeraldNet, "Many Clues to Collapse," 2007, accessed February 12, 2024, https://www.heraldnet.com/news/many-clues-to-collapse/.

10.29. America Society of Civil Engineers, "Report Card for America's Infrastructure," 2021, accessed February 13, 2024, https://infrastructurereportcard.org/wp-content/uploads/2020/12/National_IRC_2021-report.pdf.

10.30. Transport Action Network, "£20bn Backlog as Potholes Grow," 2023, accessed February 13, 2024, https://transportactionnetwork.org.uk/20bn-backlog-as-potholes-grow/.

10.31. Pundsack, M., Whapples, C. et al., "Car Park Design," The Institution of Structural Engineers, 2023, accessed February 13, 2024, https://www.istructe.org/resources/guidance/car-park-design/.

10.32. Millman, J. and Miller, M., "NYC Shutters Additional Parking Garages in Wake of Deadly Manhattan Collapse," NBC New York, 2023, accessed February 13, 2024, https://www.nbcnewyork.com/news/local/nyc-shutters-additional-parking-garages-in-wake-of-deadly-manhattan-collapse/4287611/.

10.33. Hicks, N., Marino, J., McCarthy, C., and Janoski, S., "Manhattan Garage Collapse Likely Caused by Too Many Cars on Top Floor: Fire Officials," *New York Post*, 2023, accessed February 13, 2024, https://nypost.com/2023/04/19/manhattan-da-investigating-parking-garage-collapse-caused-by-too-many-cars-on-98-year-old-buildings-top-floor/.

10.34. Renshaw, J., "As EV Sales Grow, Battle over U.S. Road Weight Limits Heats Up," Reuters, 2022, accessed February 13, 2024, https://www.reuters.com/business/autos-transportation/ev-sales-grow-battle-over-road-weight-limits-heats-up-2022-10-05/.

10.35. The Rambling Brick, "The Word on the Streets… [Vale LEGO City Road Baseplates]," 2020, accessed November 23, 2020, https://ramblingbrick.com/2020/11/23/the-word-on-the-streets-vale-lego-city-road-baseplates/.

Chapter 11

11.1. Bergek, A., Berggren, C., and K.R. Group, "The Impact of Environmental Policy Instruments on Innovation: A Review of Energy and Automotive Industry Studies," *Ecological Economics* 106 (2014): 112-123.

11.2. Stroud, D.A., "Regulation of Some Sources of Lead Poisoning: A Brief Review," in *Proceedings of the Oxford Lead Symposium*, Edward Grey Institute, University of Oxford, Oxford, UK, 2015.

11.3. National Research Council, *Effectiveness and Impact of Corporate Average Fuel Economy (CAFE) Standards* (Washington, DC: National Academies Press, 2002).

11.4. Department of Transport, "Consultation Outcome: Driving Licence Flexibility for Alternatively-Fuelled Vehicles," 2023, accessed February 15, 2024, https://www.gov.uk/government/consultations/driving-licence-flexibility-for-alternatively-fuelled-vehicles/outcome/consultation-outcome-driving-licence-flexibility-for-alternatively-fuelled-vehicles.

11.5. European Union, "UN Regulation No. 168—Uniform Provisions Concerning the Approval of Light Duty Passenger and Commercial Vehicles with Regards to Real Driving Emissions (RDE) [2024/211]," 42024X0211, European Union, Editor, 2024, accessed February 15, 2024, http://data.europa.eu/eli/reg/2024/211/oj.

11.6. Palin, R. et al., "ISO 26262 Safety Cases: Compliance and Assurance," 2011.

11.7. Kahane, C.J., "Lives Saved by Vehicle Safety Technologies and Associated Federal Motor Vehicle Safety Standards, 1960 to 2012—Passenger Cars and LTVs—With Reviews of 26 FMVSS and the Effectiveness of Their Associated Safety Technologies in Reducing Fatalities, Injuries, and Crashes," 2015.

11.8. Lie, A. and Tingvall, C., "How Do Euro NCAP Results Correlate with Real-Life Injury Risks? A Paired Comparison Study of Car-to-Car Crashes," *Traffic Injury Prevention* 3, no. 4 (2002): 288-293.

11.9. Senecal, K. and Leach, F., *Racing toward Zero: The Untold Story of Driving Green* (Warrendale: SAE International, 2021).

11.10. Jacobsen, M.R., "Fuel Economy and Safety: The Influences of Vehicle Class and Driver Behavior," *American Economic Journal: Applied Economics* 5, no. 3 (2013): 1-26.

11.11. NHTSA, "Corporate Average Fuel Economy," accessed February 16, 2024, https://www.nhtsa.gov/laws-regulations/corporate-average-fuel-economy.

11.12. Uwe Tietge, S.D., Yang, Z., and Mock, P., "From Laboratory to Road International: A Comparison of Official and Real-World Fuel Consumption and CO_2 Values for Passenger Cars in Europe, The United States, China, and Japan," ICCT, 2017, accessed February 4, 2024, https://theicct.org/publication/from-laboratory-to-road-international-a-comparison-of-official-and-real-world-fuel-consumption-and-co2-values-for-passenger-cars-in-europe-the-united-states-china-and-japan/.

11.13. Code of Federal Regulations, "Title 49 Subtitle B Chapter V Part 533 § 533.5," 2024, accessed February 16, 2024, https://www.ecfr.gov/current/title-49/subtitle-B/chapter-V/part-533/section-533.5.

11.14. Code of Federal Regulations, "Title 49 Subtitle B Chapter V Part 531 § 531.5," 2024, accessed February 16, 2024, https://www.ecfr.gov/current/title-49/subtitle-B/chapter-V/part-531/section-531.5.

11.15. Sen, B., Noori, M., and Tatari, O., "Will Corporate Average Fuel Economy (CAFE) Standard Help? Modeling CAFE's Impact on Market Share of Electric Vehicles," *Energy Policy* 109 (2017): 279-287, doi:https://doi.org/10.1016/j.enpol.2017.07.008.

11.16. Jean, B.M., "Do Footprint-Based CAFE Standards Make Car Models Bigger?," 2015.

11.17. Eisenstein, P.A., "Ford to Stop Making All Passenger Cars Except the Mustang," NBC News, 2018, accessed February 16, 2024, https://www.nbcnews.com/business/autos/ford-stop-making-all-passenger-cars-except-mustang-n869256.

11.18. European Environment Agency, "CO_2 Performance of New Passenger Cars in Europe," 2023, accessed February 9, 2024, https://www.eea.europa.eu/en/analysis/indicators/co2-performance-of-new-passenger.

11.19. Bolt, "How Cars Became Status Symbols and Why We Should Rethink That Obsession," Bolt Blog, 2023, accessed August 23, 2023, https://bolt.eu/en/blog/car-as-a-status-symbol.

11.20. Rauh, N., Franke, T., and Krems, J.F., "Understanding the Impact of Electric Vehicle Driving Experience on Range Anxiety," *Human Factors* 57, no. 1 (2015): 177-187.

11.21. EPA, "Automotive Trends Report," 2023, accessed February 13, 2024, https://www. epa.gov/automotive-trends/download-automotive-trends-report.

11.22. Lu, M., "Visualized: EV Market Share in the U.S.," Visual Capitalist, 2023, accessed February 16, 2024, https://www.visualcapitalist.com/visualized-ev-market-share-in-the-u-s/.

11.23. Ritchie, H., "The Weighty Issue of Electric Cars," 2023, accessed July 17, 2023, https:// www.sustainabilitybynumbers.com/p/weighty-issue-of-electric-cars.

11.24. Ritchie, H., "The World Likes Big Cars, the Data Don't Lie," Sustainability by Numbers, 2023, accessed February 16, 2024, https://www.sustainabilitybynumbers. com/p/global-car-economy.

Chapter 12

12.1. Gombar, V., Editor, "Electric Cars Have Dented Fuel Demand. By 2040, They'll Slash It," BloombergNEF, 2023, accessed August 15, 2023, https://about.bnef.com/blog/ electric-cars-have-dented-fuel-demand-by-2040-theyll-slash-it.

12.2. Goldman, Sachs, "Peak oil demand is still a decade away," 2024, accessed August 6, 2024, https://www.goldmansachs.com/insights/articles/peak-oil-demand-is-still-a-decade-away.

12.3. Santos, G., "Road Fuel Taxes in Europe: Do They Internalize Road Transport Externalities?" *Transport Policy* 53 (2017): 120-134, doi:https://doi.org/10.1016/j. tranpol.2016.09.009.

12.4. Masala, F., "Vehicle Excise Duty," House of Commons Library: CBP-1482, 2023, accessed February 19, 2024, https://researchbriefings.files.parliament.uk/documents/ SN01482/SN01482.pdf.

12.5. GOV.UK, "Fuel Duty," accessed February 19, 2024, https://www.gov.uk/tax-on-shopping/fuel-duty.

12.6. Office of Budget Responsibility, "Vehicle Excise Duty," 2023, accessed January 16, 2024, https://obr.uk/forecasts-in-depth/tax-by-tax-spend-by-spend/vehicle-excise-duty/.

12.7. Office of Budget Responsibility, "Fuel Duties," 2023, accessed January 16, 2024, https://obr.uk/forecasts-in-depth/tax-by-tax-spend-by-spend/fuel-duties/.

12.8. RAC Foundation, "Mobility," 2023, accessed February 19, 2024, https://www. racfoundation.org/motoring-faqs/mobility#a1.

12.9. Ritchie, H., "The Weighty Issue of Electric Cars," 2023, accessed July 17, 2023, https:// www.sustainabilitybynumbers.com/p/weighty-issue-of-electric-cars.

12.10. Green, J., "Average Miles Driven per Year by State: Why You Should Care," 2023, accessed February 19, 2024, https://www.trustedchoice.com/insurance-articles/ wheels-wings-motors/average-miles-driven-per-year/.

12.11. Carlier, M., "Automobile Registrations in the United States in 2021," State, 2023, accessed February 19, 2024, https://www.statista.com/statistics/196010/total-number-of-registered-automobiles-in-the-us-by-state/.

12.12. Newsom, G., "Governor's Budget Summary," 2023, 278, accessed February 19, 2024, https://ebudget.ca.gov/2022-23/pdf/BudgetSummary/FullBudgetSummary.pdf.

12.13. Statista, "Average Annual Passenger Car Journeys in France from 2004 to 2018," Fuel Type, 2022, accessed February 19, 2024, https://www.statista.com/statistics/1105129/distance-traveled-in-average-by-passenger-car-france/.

12.14. Eurostat, "Passenger Cars in the EU," 2023, accessed February 19, 2024, https://ec.europa.eu/eurostat/statistics-explained/index.php?title=Passenger_cars_in_the_EU.

12.15. Yanatma, S., "Which European Countries Pay the Least and Most Tax on Cars?," euronews.business, 2023, accessed February 19, 2024, https://www.euronews.com/business/2023/12/26/which-european-countries-pay-the-least-and-most-tax-on-cars.

12.16. Monteforte, M., Rajon Bernard, M., Bernard Y., Bieker G. et al., "European Vehicle Market Statistics 2022/23," ICCT, 2023, accessed February 1, 2024, https://theicct.org/publication/european-vehicle-market-statistics-2022-23/.

12.17. Bareckas, K., "What Is the Average Miles Driven per Year, and Why Is It Important?," CarVertical, 2023, accessed February 9, 2023, https://www.carvertical.com/blog/average-miles-driven-per-year.

12.18. Statista, "Average Travel Distance of Private Car Owners in Japan from 2014 to 2023," 2023, accessed February 19, 2023, https://www.statista.com/statistics/1198060/japan-average-travel-distance-private-car-owners/.

12.19. Arba, A., "Number of Passenger Cars in Use in Japan from 2014 to 2023," 2023, accessed February 19, 2024, https://www.statista.com/statistics/911570/japan-passenger-cars-in-use-numbers/.

12.20. Arba, A., "Automobile-Related Tax Income in Japan in Fiscal Year 2023, by Segment," 2024, accessed February 19, 2024, https://www.statista.com/statistics/1445191/japan-automobile-related-tax-revenue-by-segment/.

12.21. Statista, "Electricity Consumption from All Electricity Suppliers in the United Kingdom (UK) from 2000 to 2022," 2023, accessed February 19, 2024, https://www.statista.com/statistics/322874/electricity-consumption-from-all-electricity-suppliers-in-the-united-kingdom.

12.22. Statista, "Monthly Average Electricity Prices Based on Day-Ahead Baseload Contracts in Great Britain from January 2013 to December 2023," 2024, accessed February 19, 2024, https://www.statista.com/statistics/589765/average-electricity-prices-uk/.

12.23. Leape, J., "The London Congestion Charge," *Journal of Economic Perspectives* 20, no. 4 (2006): 157-176.

12.24. Goh, M., "Congestion Management and Electronic Road Pricing in Singapore," *Journal of Transport Geography* 10, no. 1 (2002): 29-38.

12.25. Borins, S.F., "Electronic Road Pricing: An Idea Whose Time May Never Come," *Transportation Research Part A: General* 22, no. 1 (1988): 37-44.

12.26. Smirti, M. et al., "Politics, Public Opinion, and Project Design in California Road Pricing," *Transportation Research Record* 1996, no. 1 (2007): 41-48.

12.27. London Assembly, "Total Cost of Greater London ULEZ Expansion," March 23, 2023, accessed February 19, 2024, https://www.london.gov.uk/who-we-are/what-london-assembly-does/questions-mayor/find-an-answer/total-cost-greater-london-ulez-expansion.

12.28. TfL, "ULEZ Income," September 20, 2023, accessed February 19, 2024, https://tfl.gov.uk/corporate/transparency/freedom-of-information/foi-request-detail?referenceId=FOI-1691-2324.

12.29. Kenis, A. and Barratt, B., "The Role of the Media in Staging Air Pollution: The Controversy on Extreme Air Pollution along Oxford Street and Other Debates on Poor Air Quality in London," *Environment and Planning C: Politics and Space* 40, no. 3 (2022): 611-628.

12.30. Clark, R., "The Backlash against Ulez Is Only the Beginning," *Spectator* 352, no. 10169 (2023): 12-14.

12.31. Raccuja, G., "Miles Better. A Distance-Based Charge to Replace Fuel Duty and VED, Collected by Insurers," Policy Exchange, 2017, accessed June 24, 2024, https://policyexchange.org.uk/wp-content/uploads/2017/07/Gergely-Raccuja-Miles-Better-Revised-Submission.pdf.

Chapter 13

13.1. Parkers, "Volkswagen Passat Specifications," accessed February 21, 2024, https://www.parkers.co.uk/volkswagen/passat/specs/.

13.2. Volkswagen, "The New Passat," accessed February 21, 2024, https://www.volkswagen.co.uk/en/new/passat.html.

13.3. Miller, C., "VW Passat Dead after 2022, Receives Special Edition Sendoff," Car and Driver, 2021, accessed June 18, 2024, https://www.caranddriver.com/news/a37069886/vw-passat-dead-2022-final-edition/.

13.4. News, F., "Volkswagen ID.7 Pro S Price and Specifications," accessed February 21, 2024, https://www.fleetnews.co.uk/electric-fleet/electric-car-and-van-data/volkswagen/id7-1840.

13.5. Volkswagen, "ID.7 | Fully Electric Fastback," accessed February 21, 2024, https://www.volkswagen.co.uk/en/electric-and-hybrid/electric-cars/id7.html.

13.6. Senecal, K. and Leach, F., *Racing toward Zero: The Untold Story of Driving Green* (Warrendale: SAE International, 2021).

13.7. Ritchie, H., "Electric Cars Are Better for the Climate than Petrol or Diesel," 2023, accessed January 26, 2023, https://www.sustainabilitybynumbers.com/p/ev-fossil-cars-climate.

13.8. SMMT, "Motorparc Vehicles in Use (UK)," 2023, accessed July 31, 2024, https://www.smmt.co.uk/vehicle-data/motorparc-vehicles-in-use-uk/.

13.9. Xu, Y., Isom, L., and Hanna, M.A., "Adding Value to Carbon Dioxide from Ethanol Fermentations," *Bioresource Technology* 101, no. 10 (2010): 3311-3319.

13.10. Finkenrath, M., "Cost and Performance of Carbon Dioxide Capture from Power Generation," 2011.

13.11. Lackner, K.S., "The Thermodynamics of Direct Air Capture of Carbon Dioxide," *Energy* 50 (2013): 38-46.

13.12. Department for Transport, "E10 Petrol Explained," 2022, accessed February 21, 2024, https://www.gov.uk/guidance/e10-petrol-explained.

13.13. Jennifer, L., "Toyota's Hydrogen Fuel Cell Vehicle Sales Saw 166% Increase," Carbon Credits, 2023, accessed November 6, 2023, https://carboncredits.com/toyotas-hydrogen-fuel-cell-vehicle-sales-saw-166-increase/.

13.14. Carlier, M., "Best-Selling Passenger Car Worldwide in 2022," 2023, accessed February 22, 2024, https://www.statista.com/statistics/239229/most-sold-car-models-worldwide/.

13.15. "Carbon Fibre in Car Production—Weighing Up the Benefits," *Just Auto Magazine*, 2020, accessed February 14, 2024, https://justauto.nridigital.com/just-auto_magazine_jun20/carbon_fibre_in_car_production_weighing_up_the_benefits.

13.16. Alcott, B., "Jevons' Paradox," *Ecological Economics* 54, no. 1 (2005): 9-21.

13.17. York, R. and McGee, J.A., "Understanding the Jevons Paradox," *Environmental Sociology* 2, no. 1 (2016): 77-87.

Chapter 14

14.1. *Oxford English Dictionary* (Oxford: Oxford University Press, 2023).

14.2. Hjelkrem, O.A. et al., "Estimation of Tank-to-Wheel Efficiency Functions Based on Type Approval Data," *Applied Energy* 276 (2020): 115463, doi:https://doi.org/10.1016/j.apenergy.2020.115463.

14.3. U.S. Department of Energy, "All-Electric Vehicles," accessed February 22, 2024, https://www.fueleconomy.gov/feg/evtech.shtml.

14.4. Gao, Y., Chen, L., and Ehsani, M., "Investigation of the Effectiveness of Regenerative Braking for EV and HEV," *SAE Transactions* 108 (1999): 3184-3190.

14.5. Bourgeois, L. et al., "EU Refinery Energy Systems and Efficiency," Concawe, March 12, 2012, accessed February 22, 2024, https://www.concawe.eu/wp-content/uploads/2017/01/rpt_12-03-2012-01520-01-e.pdf.

14.6. Gardiner, M., "Energy Requirements for Hydrogen Gas Compression and Liquefaction as Related to Vehicle Storage Needs," US Department of Energy: 9013, 2009, accessed February 22, 2024, https://www.hydrogen.energy.gov/docs/hydrogenprogramlibraries/pdfs/9013_energy_requirements_for_hydrogen_gas_compression.pdf?Status=Master.

14.7. IRENA, "Hydrogen," accessed February 22, 2024, https://www.irena.org/Energy-Transition/Technology/Hydrogen.

14.8. Mazloomi, K., Sulaiman, N.B., and Moayedi, H., "Electrical Efficiency of Electrolytic Hydrogen Production," *International Journal of Electrochemical Science* 7, no. 4 (2012): 3314-3326.

14.9. Han, J. et al., "A Comparative Assessment of Resource Efficiency in Petroleum Refining," *Fuel* 157 (2015): 292-298, doi:https://doi.org/10.1016/j.fuel.2015.03.038.

14.10. Steilen, M. and Jörissen, L., "Chapter 10—Hydrogen Conversion into Electricity and Thermal Energy by Fuel Cells: Use of H_2-Systems and Batteries," in Moseley, P.T. and Garche, J. (Eds.), *Electrochemical Energy Storage for Renewable Sources and Grid Balancing* (Amsterdam: Elsevier, 2015), 143-158.

14.11. Muradov, N., "17—Low-Carbon Production of Hydrogen from Fossil Fuels," in Subramani, V., Basile, A., and Veziroğlu, T.N. (Eds.), *Compendium of Hydrogen Energy* (Oxford: Woodhead Publishing, 2015), 489-522.

14.12. U.S. Department of Energy, "Fuel Cells," accessed February 22, 2024, https://www.energy.gov/sites/prod/files/2015/11/f27/fcto_fuel_cells_fact_sheet.pdf.

14.13. Svarc, J., "Most Efficient Solar Panels 2024," Clean Energy Reviews, 2024, accessed January 24, 2024, https://www.cleanenergyreviews.info/blog/most-efficient-solar-panels.

14.14. Energy Storage, "Hydrogen Energy Storage," accessed February 22, 2024, https://energystorage.org/why-energy-storage/technologies/hydrogen-energy-storage/.

14.15. U.S. Energy Information Administration, "Frequently Asked Questions," accessed February 22, 2024, https://www.eia.gov/tools/faqs/faq.php.

14.16. Zhu, X.-G., Long, S.P., and Ort, D.R., "Improving Photosynthetic Efficiency for Greater Yield," *Annual Review of Plant Biology* 61, no. 1 (2010): 235-261, doi:https://doi.org/10.1146/annurev-arplant-042809-112206.

14.17. Dewulf, J., Van Langenhove, H., and Van De Velde, B., "Exergy-Based Efficiency and Renewability Assessment of Biofuel Production," *Environmental Science & Technology* 39, no. 10 (2005): 3878-3882, doi:https://doi.org/10.1021/es048721b.

14.18. Ababneh, H. and Hameed, B.H., "Electrofuels as Emerging New Green Alternative Fuel: A Review of Recent Literature," *Energy Conversion and Management* 254 (2022): 115213, doi:https://doi.org/10.1016/j.enconman.2022.115213.

14.19. Grahn, M. et al., "Review of Electrofuel Feasibility—Cost and Environmental Impact," *Progress in Energy* 4, no. 3 (2022): 032010, doi:https://doi.org/10.1088/2516-1083/ac7937.

14.20. Leibenstein, H., "Allocative Efficiency vs. 'X-Efficiency'," *The American Economic Review* 56 (1966): 392-415.

14.21. Borwein, J.M., "On the Existence of Pareto Efficient Points," *Mathematics of Operations Research* 8, no. 1 (1983): 64-73.

14.22. Harrison, G.W., Lau, M.I., and Williams, M.B., "Estimating Individual Discount Rates in Denmark: A Field Experiment," *American Economic Review* 92, no. 5 (2002): 1606-1617.

14.23. Fiorello, D. et al., "Mobility Data across the EU 28 Member States: Results from an Extensive CAWI Survey," *Transportation Research Procedia* 14 (2016): 1104-1113, doi:https://doi.org/10.1016/j.trpro.2016.05.181.

14.24. Office of Energy Efficiency& Renewable Energy, "Average Vehicle Occupancy Remains Unchanged from 2009 to 2017," FOTW #1040, July 30, 2018, accessed February 22, 2024, https://www.energy.gov/eere/vehicles/articles/fotw-1040-july-30-2018-average-vehicle-occupancy-remains-unchanged-2009-2017.

14.25. Frost, N., "For 11 Years, the Soviet Union Had No Weekends," September 7, 2023, accessed February 22, 2024, https://www.history.com/news/soviet-union-stalin-weekend-labor-policy.

Chapter 15

15.1. Adda, J. and Cornaglia, F., "The Effect of Bans and Taxes on Passive Smoking," *American Economic Journal: Applied Economics* 2, no. 1 (2010): 1-32.

15.2. Senecal, K. and Leach, F., *Racing toward Zero: The Untold Story of Driving Green* (Warrendale: SAE International, 2021).

15.3. Corning, "What Emissions Regulations and Standards Are Currently in Place?," accessed February 9, 2024, https://www.corning.com/emea/en/products/environmental-technologies/emissions-control/emissions-regulations.html.

15.4. Stone, R., *Introduction to Internal Combustion Engines*. 3rd Ed. (London: Springer, 1999).

15.5. Williams, M. and Minjares, R., "A Technical Summary of Euro 6/VI Vehicle Emission Standards," ICCT, 2016, accessed February 1, 2024, https://theicct.org/sites/default/files/publications/ICCT_Euro6-VI_briefing_jun2016.pdf.

15.6. Burton, T. et al., "A Data-Driven Greenhouse Gas Emission Rate Analysis for Vehicle Comparisons," *SAE Int. J. Elec. Veh.* 12, no. 1 (2023): 91-127, doi:https://doi.org/10.4271/14-12-01-0006.

15.7. Chen, S. et al., "A Critical Review on Deployment Planning and Risk Analysis of Carbon Capture, Utilization, and Storage (CCUS) toward Carbon Neutrality," *Renewable and Sustainable Energy Reviews* 167 (2022): 112537.

15.8. Singh, A. et al., "Impacts of Emergency Health Protection Measures upon Air Quality, Traffic and Public Health: Evidence from Oxford, UK," *Environmental Pollution* 293 (2022): 118584, doi:https://doi.org/10.1016/j.envpol.2021.118584.

15.9. European Parliament and the Council of the European Union, "Regulation (EC) No. 715/2007 of the European Parliament and of the Council of 20 June 2007 on Type Approval of Motor Vehicles with Respect to Emissions from Light Passenger and Commercial Vehicles (Euro 5 and Euro 6) and on Access to Vehicle Repair and Maintenance Information (Text with EEA relevance), in 32007R0715," *Official Journal of the European Union*, 2007, 16, accessed February 9, 2024, https://eur-lex.europa.eu/legal-content/EN/TXT/?uri=CELEX%3A32007R0715&qid=1707497731670.

15.10. GerMan, J., "Estimated Cost of Emission Reduction Technologies for Light-Duty Vehicles," The International Council on Clean Transportation, Washington, DC, 2012.

15.11. Justia, "USA, et al. v. VOLVO POWERTRAIN CORPORATION," United States District Court for the District of Columbia," 2011, https://law.justia.com/cases/federal/district-courts/district-of-columbia/dcdce/1:1998cv02547/69130/67/.

15.12. Thompson, G.J. et al., "In-Use Emissions Testing of Light-Duty Diesel Vehicles in the United States," 2014.

15.13. European Court of Auditors, "Special Report 01/2024: Reducing Carbon Dioxide Emissions from Passenger Cars—Finally Picking up Pace, but Challenges on the Road Ahead," 2024, accessed February 9, 2024, https://www.eca.europa.eu/en/publications/SR-2024-01.

15.14. European Parliament, "Euro 7: Deal on New EU Rules to Reduce Road Transport Emissions," 2023, accessed February 9, 2024, https://www.europarl.europa.eu/news/en/press-room/20231207IPR15740/euro-7-deal-on-new-eu-rules-to-reduce-road-transport-emissions.

15.15. Suarez, J. et al., "2025 and 2030 CO_2 Emission Targets for Light Duty Vehicles," *Publ. Off. Eur. Union* 10 (2023): 901734.

15.16. European Environment Agency, "CO_2 Performance of New Passenger Cars in Europe," 2023, accessed February 9, 2024, https://www.eea.europa.eu/en/analysis/indicators/co2-performance-of-new-passenger.

15.17. Marotta, A. et al., "Gaseous Emissions from Light-Duty Vehicles: Moving from NEDC to the New WLTP Test Procedure," *Environmental Science & Technology* 49, no. 14 (2015): 8315-8322.

15.18. Uwe Tietge, S.D., Yang, Z., and Mock, P., "From Laboratory to Road International: A Comparison of Official and Real-World Fuel Consumption and CO_2 Values for Passenger Cars in Europe, The United States, China, and Japan," ICCT, 2017, accessed February 4, 2024, https://theicct.org/publication/from-laboratory-to-road-international-a-comparison-of-official-and-real-world-fuel-consumption-and-co2-values-for-passenger-cars-in-europe-the-united-states-china-and-japan/.

15.19. European Commission, "CO_2 Emission Performance Standards for Cars and Vans," accessed February 9, 2024, https://climate.ec.europa.eu/eu-action/transport/road-transport-reducing-co2-emissions-vehicles/co2-emission-performance-standards-cars-and-vans_en.

15.20. National Research Council, *Effectiveness and Impact of Corporate Average Fuel Economy (CAFE) Standards* (Washington, DC: National Academies Press, 2002).

15.21. California Air Resources Board, "California Moves to Accelerate to 100% New Zero-Emission Vehicle Sales by 2035," 2022, accessed February 9, 2024, https://ww2.arb.ca.gov/news/california-moves-accelerate-100-new-zero-emission-vehicle-sales-2035.

Chapter 16

16.1. Strzelec, A. and Kasab, J., *Automotive Emissions Regulations and Exhaust Aftertreatment Systems* (Warrendale: SAE International, 2020).

16.2. Senecal, K. and Leach, F., *Racing toward Zero: The Untold Story of Driving Green* (Warrendale: SAE International, 2021).

16.3. Ravi, S.S. et al., "On the Pursuit of Emissions-Free Clean Mobility – Electric Vehicles versus e-Fuels," *Science of the Total Environment* 875 (2023): 162688, doi:https://doi.org/10.1016/j.scitotenv.2023.162688.

16.4. The Economist, "Western Firms Are Quaking as China's Electric-Car Industry Speeds Up," 2024, accessed February 9, 2024, https://www.economist.com/briefing/2024/01/11/western-firms-are-quaking-as-chinas-electric-car-industry-speeds-up.

16.5. European Council, "'Fit for 55': Council Adopts Regulation on CO_2 Emissions for New Cars and Vans," 2023, https://www.consilium.europa.eu/en/press/press-releases/2023/03/28/fit-for-55-council-adopts-regulation-on-co2-emissions-for-new-cars-and-vans/.

16.6. European Parliament, "Euro 7: Deal on New EU Rules to Reduce Road Transport Emissions," 2023, accessed February 9, 2024, https://www.europarl.europa.eu/news/en/press-room/20231207IPR15740/euro-7-deal-on-new-eu-rules-to-reduce-road-transport-emissions.

16.7. Braisher, M., Stone, R., and Price, P., "Particle Number Emissions from a Range of European Vehicles," SAE Technical Paper 2010-01-0786 (2010), doi:https://doi.org/10.4271/2010-01-0786.

16.8. Reuters, "Germany to End e-Vehicle Subsidy Programme," 2023, accessed February 9, 2024, https://www.reuters.com/business/autos-transportation/germany-end-e-vehicle-subsidy-programme-2023-12-16/.

16.9. Lipsey, R.G. and Lancaster, K., "The General Theory of Second Best," *The Review of Economic Studies* 24, no. 1 (1956): 11-32, doi:https://doi.org/10.2307/2296233.

List of Abbreviations

6PPD	N^1-(4-Methylpentan-2-yl)-N^4-phenylbenzene-1,4-diamine
ACEA	European Automobile Manufacturers' Association
ADAS	Advanced Driver-Assistance System
ADP	Abiotic Depletion Potential
BEV	Battery Electric Vehicle
CAFE	Corporate Average Fuel Economy
CARB	California Air Resources Board
CCGT	Combined Cycle Gas Turbines
CCUS	Carbon Capture, Utilization, and Storage
CFRP	Carbon Fiber-Reinforced Plastic
DPF	Diesel Particulate Filter
DVLA	Driver and Vehicle Licensing Agency
EPA	Environmental Protection Agency (US Federal)
ERP	Electronic Road Pricing
ETRMA	European Tyre and Rubber Manufacturers' Association
EU	European Union
F1	Formula 1
FCEV	Fuel Cell Electric Vehicle
FDP	Fossil Depletion Potential
FETP	Freshwater Eco-Toxicity Potential
FMVSS	Federal Motor Vehicle Safety Standards
GDI	Gasoline Direct Injection

GPF	Gasoline Particulate Filter
GREET	Greenhouse Gases, Regulated Emissions, and Energy Use in Transportation Model
GVWR	Gross Vehicle Weight Ratings
GWP	Global Warming Potential
HOV	High-Occupancy Vehicle
HTP	Human Toxicity Potential
ICCT	International Council on Clean Transportation
ICE	Internal Combustion Engine
ICEV	Internal Combustion Engine Vehicle
IFPEN	IFP Energies Nouvelles
IRV	Impact Reduction Values
ISO	International Organization for Standardization
LCA	Life Cycle Assessment
LCFS	Low Carbon Fuel Standard
LFP	Lithium Iron Phosphate
LPG	Liquified Petroleum Gas
MDP	Mineral Depletion Potential
mpg	miles per gallon
NCAP	New Car Assessment Program (or Programme)
NEDC	New European Driving Cycle
NHTSA	National Highway Traffic Safety Administration
NO$_x$	Nitrogen Oxides
OBR	Office for Budget Responsibility
OECD	Organisation for Economic Co-operation and Development
PN	Particle Number
PSA	Peugeot Société Anonyme
RDE	Real Driving Emissions
RUP	Road User Pricing
SCR	Selective Catalytic Reduction

SMR	Steam Methane Reformation
SUV	Sports Utility Vehicle
TAP	Terrestrial Acidification Potential
TCJA	Tax Cuts and Jobs Act
TCO	Total Cost of Ownership
TETP	Terrestrial Eco-Toxicity Potential
UK	United Kingdom
ULEZ	Ultra Low Emissions Zone
US	United States
VAT	Value Added Tax
VED	Vehicle Excise Duty
VOC	Volatile Organic Compounds
WLTP	Worldwide Harmonised Light Vehicle Test Procedure
WTW	Well-to-Wheel
ZEV	Zero-Emission Vehicle
ZLEV	Zero- and Low-Emission Vehicle

Index